ADHESION 13

This volume is based on papers presented at the 26th annual conference on Adhesion and Adhesives held at The City University, London

Previous conferences have been published under the titles of
Adhesion 1–12

ADHESION 13

Edited by

K. W. ALLEN

Adhesion Science Group,
The City University, London, UK

ELSEVIER APPLIED SCIENCE
LONDON and NEW YORK

ELSEVIER SCIENCE PUBLISHERS LTD
Crown House, Linton Road, Barking, Essex IG11 8JU, England

Sole Distributor in the USA and Canada
ELSEVIER SCIENCE PUBLISHING CO., INC.
655 Avenue of the Americas, New York, NY 10010, USA

WITH 38 TABLES AND 179 ILLUSTRATIONS

softcover reprint of the hardcover 1st edition 1989

British Library Cataloguing in Publication Data
Conference on Adhesion and Adhesives (26th: 1988: City
    University).
    Adhesion 13
    1. Adhesion
    I. Title   II. Allen, K.W.
    541.3'453

Library of Congress Cataloging in Publication Data
Conference on Adhesion and Adhesives (26th: 1988: City
    University, London)
    Adhesion 13.

    'This volume is based on papers presented at the 26th Annual
Conference on Adhesion and Adhesives held at the City
University, London.'
    1. Adhesion—Congresses.   2. Adhesives—Congresses.
I. Allen, K. W.   II. Title.
TP967.C64   1988   668'.3   88-33587

ISBN-13: 978-94-010-9084-1   e-ISBN-13: 978-94-010-9082-7
DOI: 10.1007/978-94-010-9082-7

Preface

As I turn to write this Preface to another volume of papers presented at the Annual Conference on Adhesion and Adhesives, I am reflecting on how fortunate we are to be able to continue to attract both speakers and audience of such high standard year after year. This field of science and technology continues to grow in size and importance, so that it is not so surprising that those interested from a very considerable range of countries and disciplines are willing to come together and spend a couple of days each year exchanging ideas and views, both formally and informally.

So, once again, I can and must express thanks to all those who have made this volume possible: both those who presented the papers, and then rewrote them for publication, and those who through their attendance made the conference a viable reality.

K. W. ALLEN

Contents

viii

1

THE U.S. ARMY'S ADHESIVE BONDING IMPROVEMENT INITIATIVE

STANLEY E. WENTWORTH
Research Chemist/Polymer Research Branch
U.S. Army Materials Technology Laboratory
Watertown, MA 02172-0001 USA

INTRODUCTION

During 1985, an incident involving the debond in flight of the rain
erosion boot from the blade of a Cobra helicopter called attention to
adhesive bonding problems in Army aircraft. A study of such problems,
later expanded to include all categories of Army materiel, identified
the root cause of these problems and recommended a course of action
to reduce or eliminate such problems in current systems and prevent
their occurence in systems now in development or on the drawing board.
The balance of this paper presents the results of this study and its
recommendations.

CURRENT PROBLEMS

The study brought to light a very large number of bonding problems
ranging in criticality from largely cosmetic issues such as the bonding of a data plate to a vehicle dashboard to basic mission capability
issues such as the integrity of primary structural bonds in missile
applications.
   In virtually every case examined, the cause of the bonding
problem was found not to be the result of any fundamental lack of
knowledge but rather the failure to treat adhesive bonding as a
serious structural issue. Very often, this led to inadequate control
and/or specification of the bonding process. In some cases, improper
surface treatment prior to bonding was responsible for poor joint
performance. For example, the data plate referred to above was bonded
to a polyurethane painted surface without any surface preparation.
In other cases, the bonding operation was conducted under less than
ideal conditions. For instance, in one case improper humidity control
resulted in excessive moisture uptake by a substrate material.
   Another major consequence of the failure to treat bonding seriously
is that the adhesive joint is not integrated into the total structural
design of the system. The joint is therefore not optimized for its
structural role and may indeed be unable to serve its intended purpose.
Correlary to this is the failure to consider the end-use environment
in which the bonded structure will operate. This can lead to the

selection of an inadequate adhesive. Bonds often fail in service for this reason alone.

Quality control is another area where lack of attention to detail causes problems. Often, there is no requirement to monitor batch to batch variation in the composition of adhesives. Yet it is well known that such variations can lead to bonds of differing strength and durability.

Finally, the nondestructive evaluation (NDE) of adhesive joints was identified as an area where new knowledge is required to meet short term deficiencies. Specifically, no nondestructive method exists which can reliably indicate the strength of an adhesive joint. According to the "if you can't inspect it, you can't fly it" philosophy, this deficiency can be seen to be a major barrier to the use of adhesive bonding in primary structure.

A number of steps have been taken to correct the deficiencies found in adhesive bonding in current Army materiel. Firstly, specifications and standards dealing with adhesive bonding have been reviewed and are being updated to reflect the state-of-the-art in adhesive bonding technology. By this means, contract documents will insure that the best available bonding methodology is used in the manufacture of Army equipment. The recently issued update of Military Standardization Handbook 691B "Adhesive Bonding" is an especially good example of this process [1].

Secondly, the adhesive data base which has been maintained at Picatinny Arsenal for several years, is being augmented and expanded. Not only will data on new adhesives and new data on existing adhesives be added, but a new feature entitled Lessons Learned is being incorporated. Thus, descriptions of bonding problems and their solutions will be retrievable. This data base will be a valuable resource to those seeking to utilize adhesive bonding in new applications as well as to those attempting to solve current bonding problems.

Another step is being taken in the area of NDE. Here, several activities are under way. A state-of-the-art review of NDE of adhesive joints has been conducted and is about to be published and a major Government/Industry Workshop on adhesive bond NDE was held in the spring of 1988. This workshop brought to light some very promising approaches to bond strength NDE. Other activities include an evaluation of the utility of advanced signal analysis in conjunction with ultrasonic spectroscopy and the initiation of a Small Business Innovative Research program dealing with novel approaches to NDE.

Perhaps the major step in addressing current problems, however, is the development of a model approach to their solution. Accordingly, each major subordinate command (e.g., Missile Command, Tank-Automotive Command) has been tasked with selecting a bonding problem for which a solution is to be sought. Each of these problems will be analyzed in terms of the shortcomings identified above. Appropriate corrective actions such as better surface treatment, better process control and/or adhesive replacement will be taken. Close coordination between various commands and Army laboratories will highlight this process. Each command will then apply this same methodology, tailored to its needs, to the solution of other bonding problems within its purview. Results of this program will be entered into the data base as detailed above.

Finally, as a means of enhancing communications and alerting academia and industry to the Army's needs in the area of adhesive bonding, a Government/Industry Symposium on Structural Adhesive Bonding was held at Picatinny Arsenal, Dover, NJ, in early November of 1987, and the Army's annual Sagamore Materials Research Conference for 1988 had Adhesive Bonding Science and Technology as its theme.

FUTURE NEEDS

Future needs or long range problems are those whose solutions require the generation of new knowledge. This new knowledge is the product of the current and planned R & D program in adhesion science and technology.

At this point it is useful to recognize the interdisciplinary nature of adhesion science and that true progress can come only through a consideration of the interrelationship between chemistry, mechanics and surface science for both adhesive and adherend.

Enormous strides have been made in each of these fields over the past two decades. In chemistry, for instance, a good understanding of polymer structure/property relationships is now at hand such that the optimization of a resin for a specific application can now proceed along rational lines. Analogous advances have been made in mechanics where the availability of enormous computing power at very reasonable cost has lead to the development of refined computer codes for the prediction of the response of bonded joints under a wide range of conditions.

The most dramatic advances, however, have occurred in the area of surface science. The development of major new instrumental techniques for surface analysis and their application to the study of failed adhesive joints has provided valuable insight into mechanisms by which failures occur for many classes of materials under many conditions. Such information is, of course, essential for the reduction and elimination of such failures.

In spite of these advances, much remains to be done. In the area of chemistry, there is a need for improved adhesives in terms of strength and resistance to extreme environments. In addition, new materials such as metal matrix composites may require new surface treatments, a process often involving chemical modification of the surface. Surface analytic techniques are required in order to understand the nature of these transformations. Other areas of surface science needing attention are methods for studying the "interphase" and an examination of the role of ion implantation in the preparation of surfaces for adhesive bonding. In the area of mechanics, development of more comprehensive computer codes for the prediction of adhesive joint behavior including long term performance is required. In addition, work is needed to elucidate the response of bonded joints to impact loading as is work on the statistics of adhesive joint failure and reliability. A more detailed discussion of each of these areas follows below.

Research Needs in Chemistry

Even though there is currently a large number of commercially available adhesives suitable for a wide range of applications, the need exists in Army systems now in development for adhesives with enhanced performance over these state-of-the-art materials. In lightweight bridging for example, adhesives having the following attributes are needed:

Long shelf life at ambient temperature.

Service life of 15 years.

Lap shear strength of 10,000 psi.

In addition, the need exists for adhesives with substantially higher upper use temperatures. Such materials would be utilized in composite gun tubes, missile structures and in or near aircraft power plants.

The basic approach to all of these advances is through synthetic modification of the adhesive base resin along lines indicated by advances in polymer science over the past years. Optimization for specific applications would be accomplished through formulation. A series of test and evaluation studies would be required to validate such a new material for a specific system.

Modification of adherend surfaces is usually achieved by chemical means and is thus considered in this section on chemistry although it could equally well be discussed under surface science. Indeed, surface modification by ion implantation will be discussed in that section.

The importance of the role of adherend surface preparation in securing strong, durable adhesive bonds cannot be overstated. Because of its importance to the aerospace industry, the surface preparation of aircraft aluminum is sufficiently well understood that when done properly, very reliable bonds are routinely produced. In this case, a very strong, well characterized surface oxide layer is produced by means of an electrochemical etching process. Recent work has shown that this surface can be even further enhanced thru the addition of hydration inhibitors which retard the moisture assisted transformation of the oxide initially formed to a less mechanically strong crystal modification [2]. Ideally, this level of understanding should be at hand for any material to be adhesively bonded.

Work needed in this area falls into two major categories: new surface preparations and surface preparations for new materials. While, as indicated, aircraft aluminum appears to be well in hand, the same is not the case for other materials. There are still major problems in bonding titanium which appear to be related to the lack of a proper surface treatment. The same is true for steel and for several organic materials including polyolefins, various rubbers and resin matrix composites. Each of these materials requires its own individual approach based on a recognition of its surface chemistry. Methods to roughen the surface are important but so are methods which modify the surface chemical composition. For each treatment developed, a test and evaluation program to establish the strength and durability of bonds to the resultant surface must be conducted.

As new materials now in development are proposed for introduction into Army systems, surface treatments for their joining by adhesive bonding must be developed. Examples of such materials are:

Metal Matrix Composites.

Thermoplastic Matrix Composites.

Lithium/Aluminum Alloys.

Work is already underway on the first two of these materials.

Preliminary results from our laboratory on the surface treatment of a silicon carbide particulate reinforced aluminum indicate significant surface enrichment by silicon carbide. We expect but have not yet demonstrated that the strength and durability of bonds to this surface will differ significantly from those to conventional aluminum.

A program to evaluate the bondability of thermoplastic matrix composites is underway at Imperial College, London. The first phase of this program is now complete. It was shown that standard surface treatments

such as abrasion, solvent wipe and acid etch did not produce bondable
surfaces on PEEK matrix composites. Corona discharge in air however did
result in surfaces to which very acceptable bonds could be formed. Work
now underway is addressing the nature of the surfaces produced by corona
discharge [3].

Research Needs in Surface Science

It is now generally conceded that the practice of adhesive bonding of
structural elements is no longer an art. Rather, it has achieved the
status of an advanced technology with a sound scientific base. Far and
away the most important factor in this evolution has been the major advances
made in the field of surface science. Of particular significance has been
the development of instrumentation which permits very high resolution
analysis of surface chemistry and morphology.

In spite of these advances, there are still opportunities for surface
science to contribute further to progress in adhesion science and technology.
At present, there are several instrumental techniques available for surface
analysis. These are:

Scanning Electron Microscopy.

Transmission Electron Microscopy.

X-Ray Photoelectron Spectroscopy.

Scanning Auger Spectroscopy.

Energy-Despersive X-Ray Analysis.

Each of these has its own strengths and weaknesses and it is fair to say
that there are aspects of surface analysis where deficiencies still exist.
Thus, as new discoveries in physics lead to the development of new instrumental capabilities, they should be applied to adhesion science
problems as soon as practicable.

Surface modification by ion implantation is an area that has just
barely been addressed. Published work has shown that a platinum surface
can be converted from one to which no bonding could be obtained to one
exhibiting excellent bondability [4]. Obviously, this is a fertile field
for further exploration. An especially attractive feature of ion implantation is the exquisite degree to which the surface can be controlled. It is
possible to modify only the outermost layer of a material. It is also
possible to control the chemistry to a degree not attainable by conventional
means. This is a consequence of the fundamental nature of the active agent
which may be individual atoms. While this field has great promise, it
must also be regarded as high risk. A tremendous amount of work is required
to develop the parameters for a given surface and to validate the efficacy
of bonds to that surface. In addition, the development of ion implantation
equipment of sufficient size and simplicity for use in production is also a
significant barrier to practical implementation. Nevertheless, the payoff
would seem to justify the investment of sufficient resources to unequivocally
ascertain the feasability of preparing surfaces for bonding by this means.

Micromechanics is a methodology currently being utilized in the
microelectronics field. In essence, it consists of the measurement of
mechanical properties of materials while simultaneously observing the test

specimen by means of a surface analytic technique such as scanning electron microscopy. This very powerful combined technique should have great value for the study of adhesive joint behavior. It should be possible for instance, to follow fatigue crack propagation in real time and so better understand the details of the process. This would be a very costly area to pursue. However, the cost/benefit ratio favors the pursuit of this capability, at least to the point of establishing feasability.

The final area needing attention to be discussed in this section is the development of techniques for the study of the "interphase." The interphase is defined as the region of an adhesive joint including the interface between the adhesive layer and the adherend as well as the bulk of the adhesive and a representative depth into the bulk of the adherend. It includes all of the adherend oxide layer for any material which forms such a layer. This region is truly "where it's happening" for the great bulk of adhesive joints which do not fail cohesively in the adherend. The chemical, physicochemical and physical (mechanical) bonds formed across this region totally control the strength and durability of the resultant adhesive joint. This is an extremely complex region. The surface oxide layer may, for instance, cause the segregation of adhesive constituents in this region due to its surface activity of the sort that makes such oxides useful absorbants in chromatography. Thus, the chemical composition and consequently, the structure of the cured adhesive may be radically different in this crucial region. Analytic techniques are needed to probe for this sort of inhomogeneity in the in situ joint. This should be done non-destructively so that the measurement does not perturb the sample. Afterwards, the bond can be tested to destruction and a correlation made between the composition so determined and the resultant mechanical properties. This sort of information would also be invaluable for the establishment of non-destructive test methods. Thus, the development and application of methods for interrogating the interphase are essential for the understanding of adhesive joint behavior.

Research Needs in Mechanics

Since the ultimate goal in forming an adhesive joint is to form a mechanically strong union between two adherends, the availability of scientifically sound guidance for the design of such joints and of meaningful methods to test their performance is a critical need. It is the role of mechanics to provide these capabilities.

There have been major advances in this field over the last 15 years, many of them due to the tremendous increase in the availability in computing power. Much remains to be done however, especially in areas where the Army has its own special needs. For example, helicopters impose much more severe fatigue loads on joints than do fixed wing aircraft due to the vibrations generated during operation. The Army must also be more concerned with the effect of high strain rate loading (ballistic impact) on the integrity of adhesive joints, in part because of it's lead agency status in the armor field. Application of adhesive bonding to tank-automotive and armament systems also gives rise to concern for the bonding of thick structures rather than the thin materials encountered in aviation systems.

In the design area, there is a need for more powerful computer codes in order to more accurately predict the loads on and behavior of adhesive joints. Further refinement of finite element codes could incorporate durability factors such as diffusion rates for moisture and kinetic expressions for degradation reactions. Ultimately, a model for an entire life cycle is possible. This would be of great value in joint design.

Recent advances in physics have lead to the development of Laser Speckle Interferometry, a valuable tool for the study of the deformation of materials under load. Application of this technique to adhesively bonded structures will provide useful insight into their behavior and mode of failure under a variety of load conditions, again useful information for joint design. The method could also form the basis of a nondestructive test method.

Finally, the need exists for better test methods for the evaluation of the strength and durability of adhesive joints. For instance, there is much debate over the value of the lap shear test for the determination of adhesive strength. There are examples in which the relative ranking of a series of adhesives by the lap shear test was subsequently shown to be in error by more exhaustive evaluation. The lap shear test is used because it is simple. There is a need for a more valid test of equal simplicity.

SUMMARY AND CONCLUSIONS

A study of bonding problems in Army materiel has revealed that the great bulk of such problems arise not from a deficiency in the state-of-the-art in adhesive bonding but from a failure to utilize the state-of-the-art. This, in turn, is the result of the failure to treat adhesive bonding as a serious structural issue in either the design or the materials selection and process control phases of systems development and manufacture. A plan of action aimed at correcting these problems has been formulated and is being implemented.

So that similar problems can be avoided in future or "notional" systems, a research program addressing chemistry, mechanics and surface science has also been initiated. Its long term objectives are the enhancement of adhesive formulations, the development of methods for bond strength NDE and, ultimately, the development of computer models for adhesive joints which can be used not only for design but also for service life prediction under a broad range of end-use conditions. These objectives can only be realized if the multidisciplinary nature of adhesion science is recognized and the practioners of the various subdisciplines learn to speak and to understand each others language.

REFERENCES

1. MIL-HDBK 691B, Adhesive Bonding, 12 March 1987.
2. Hardwick, D. A., Ahearn, J. S. and Venables, J. D., Environmental Durability of Aluminum Adhesive Joints Protected with Hydration Inhibitors, J. Materials Sci., 19, 1984, p. 223.
3. Kodokian, G. K. A. and Kinloch, A. J., Surface Pretreatment and Adhesion of Thermoplastic Fibre-Composites. J. Materials Sci. Letters, in press.
4. Hale, E. B., James, W. J., Sharma, A. K., and Yasuda, H. K., Use of Ion Implantation to Improve Adhesion Between a Polymer Coating and a Metal, Proceed. 3rd Internat. Conf. on Mod. of Surface Props. of Metals by Ion Implant., 1981, p. 167.

2

THE SURFACE CHARACTERISATION AND ADHESIVE BONDING OF ALUMINIUM

R.J. DAVIES and A.J. KINLOCH
Department of Mechanical Engineering, Imperial College of Science and Technology,
Exhibition Rd., London, SW7 2BX.

INTRODUCTION

The adhesive bonding of aluminium alloys is of considerable interest to the aerospace industry and the satisfactory performance of such adhesive joints throughout the required service-life of the bonded component is of major importance. To achieve good resistance to environmental attack, especially by moisture, it is well established [1-3] that some form of surface pretreatment of the aluminium alloy is essential. Such pretreatments as commonly used by industry have been developed empirically. However, this empirical approach to the development of surface pretreatments has led to a lack of firm understanding as to why some treatments impart superior durability to the bonded joint than others.

The present study is investigating in detail the nature of the oxide formed on a aluminium-magnesium alloy which has been pretreated using the common methods employed by the aerospace industry. Further, the interactions between the oxide and the primer employed are being assessed. The results from these studies are also being correlated to novel durability tests, based upon a fracture mechanics approach, and to the nature of the fracture surfaces after environmental attack. The full details of the experimental methods and results will be given in a later publication and the present paper will basically review some of the initial results from this study and offer some observations on the likely mechanisms of environmental attack.

8

RESULTS and DISCUSSIONS

Materials

The aluminium-magnesium alloy employed was to British Standard 5251 and had a nominal composition of 2.25% magnesium, 0.25% manganese with the balance being aluminium.

The surface pretreaments studies included:

(i) A chromic-acid etch (CAE); according to USA Forest Products Laboratory (FPL) specification.

(ii) A chromic-acid anodise (CAA); according to UK Ministry of Defence standard DEF STAN 03-24/1.

(iii) A phosphoric-acid anodise (PAA); according to USA Boeing Aircraft Corporation specification BAC 5555.

(iv) A sulphuric-acid anodise (SAA); according to UK Ministry of Defence standard DEF STAN 03-26/1.

(v) A sulphuric-acid anodise/phosphoric-acid dip (SAA/PAD).

(Full details of the above treatments may be found in reference [4].)

A hot-cured primer was employed in most of the work reported in the present paper and was an epoxy-phenolic based primer (BR127, supplied by Cyanamid, USA). The adhesive was an unmodified single-part epoxy-paste adhesive (AV 1566 supplied by Ciba Geigy, UK) cured for one hour at 150°C.

Characterisation Studies

The main results from these characterisation studies are shown in Table I and are discussed in detail below.

TABLE I
General properties of oxide produced by the pretreatment of aluminium.

Treatment	Thickness of oxide (μm)	Type of oxide (pore size)	Structure	XPS Data	Hydration Resistance*
CAE	0.07	Open	Amorphous	no Cr	15 minutes
CAA	3.5	Partly sealed (variable)	Amorphous	no Cr	30 minutes
PAA	0.5	Open (40 nm)	Amorphous	P	16 hours
SAA	10 to 12	Sealed (15 nm)	Amorphous	S	1 hour
SAA/PAD	6 to 8	Partly sealed (15 nm)	Amorphous	P + S	3 hours

* The time to first signs of hydration in distilled water at $55^{\circ}C$; ascertained using SEM.

Surface Characterisation

The surfaces of the pretreated aluminium alloy were characterised using several complimentary methods. The main techniques employed were high-resolution scanning electron microscopy, X-ray diffraction (XRD) and X-ray photoelectron spectroscopy (XPS). Scanning electron micrographs of the various oxides produced by the different pretreatments are shown in Figures 1 to 5.

The oxide produced by the chromic-acid etch (CAE) is shown in Figure 1 and is relatively thin, being approximately 70 nm thick, and consists of closely packed clumps of deposited oxide 80 to 130 nm wide. Although the etch is essentially a chromate solution, XPS revealed that no chromium is present in the oxide. Hence, no chromates are present in the oxide layer to inhibit any subsequent hydration processes, as has been suppposed in previous work. XRD showed the oxide to amorphous in character.

The European aerospace industry prefers to use a chromic-acid etch followed by chromic acid anodising, producing a 3.5 μm thick porous oxide (figure 3), which offers better corrosion protection of the aluminium surface away from any adhesively-bonded area. This oxide essentially produces a "branch-like" porous morphology with the dimensions of the pores varying through the thickness of the oxide layer as a result of the changing anodising conditions, i.e voltage, during pretreatment. At the surface of the oxide the pore size is at its smallest (approximately 15 nm) and so a partly-closed structure is produced. XPS again revealed that no chromates are present in the oxide and XRD showed that the oxide was amorphous.

If following the CAE the aluminium is phosphoric-acid anodised, a 0.5 μm thick porous oxide with a distinctly "columnar-like" topography and open pore structure is produced (Figure 2). The pores are about 40 nm in size and appear to run continuously from the surface of the oxide to the underlying barrier layer. This oxide is found to contain phosphates which has been reported to improve the resistance to hydration of the oxide (6). XRD showed the oxide to be amorphous in character. The PAA pretreatment was developed by the Boeing Aircraft Corporation, USA.

Sulphuric acid anodising (SAA) is used to produce a relatively thick, dense oxide consisting of a close packed "drinking straw" structure with a pore size of approximately 15nm. However, a 400 nm thick hydroxide "crust" exists on the surface which effectively seals the oxide offering excellent corrosion protection. However, on the other hand, this aspect is thought to lead to the poor adhesive bonding performance of the SAA pretreatment. Indeed, the adhesive bonding performance of this oxide can be improved by the phosphoric-acid dip (SAA/PAD) treatment which removes this hydroxide crust and can produce a distinctive "brush-like" structure (figure 5). In the case of the SAA oxide, the XPS studies revealed the presence of sulphates in the oxide and XRD showed that the oxide was amorphous. In the case of the SAA/PAD oxide the XPS work also revealed the presence of phosphates in the oxide and XRD showed that the oxide was amorphous.

12

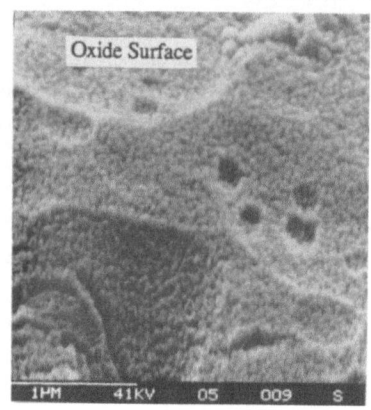

Figure 1 Topography of the oxide produced by
the FPL etch

Figure 2 Cross section and topography
of PAA oxide.

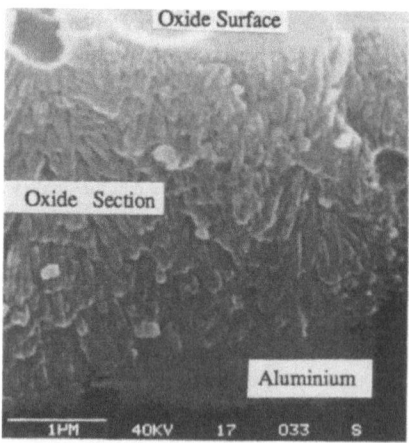

Figure 3 Cross section and topography of CAA oxide.

 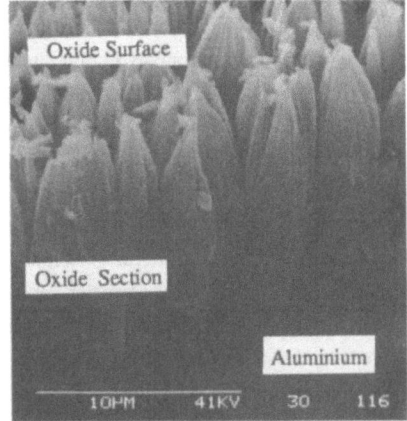

Figure 4 Cross section and topography of a thin SAA oxide.

Figure 5 Cross section and topography of a SAA oxide after the phosphoric acid dip.

Hydration Resistance of the Oxides

The hydration resistance of the various free oxide surfaces was assessed by immersing the pretreated aluminium surface in water at $55^{\circ}C$ and after various periods of time removing the samples and viewing the oxide using high resolution scanning electron microscopy. The surface and cross section of the oxide was then assessed for any morphological changes due to hydration. It has been found that three distinct stages are observed during the reaction with water for all the oxides, the differences arising from the different pretreatments being only the individual time periods between and during stages.

First an induction period is observed in which no hydroxide formation takes place, and this is probably limited by both the resistance of the oxide to surface hydrolysis and the diffusion of soluble products. In the second stage, nucleation and growth of the hydroxide occurs at an accelerating rate. In the third stage the transport of soluble species through the hydroxide layer controls the rate of hydroxide growth and the rate of hydration decreases markedly. Each of these stages are accompanied by a characteristic change in oxide morphology and volume as observed in previous work [5]. For example, once hydration begins small "bobbles" of hydroxide can be observed on the surface which then develop into protrusions and finally by dissolution and repreciation form a random pile up of loosely connected "platelets" or "cornflakes" at the hydroxide/water interface (see Figure 6). It has been suggested by Davis and Venables [6] that failure in adhesively-bonded aluminium joints is initiated as a consequence of the formation of such a weak hydroxide; the weak hydoxide now, of course, being formed *in-situ* as water permeates into the joint.

The results from our hydration studies for the exposed CAE and PAA oxides are displayed in Figure 7 as a plot of total thickness (i.e thickness of any hydroxide present and any remaining oxide) and hydroxide thickness versus time of immersion in distilled water at $55^{\circ}C$. The exceptional hydration characteristics of the PAA oxide which has been reported previously [7] is confirmed by the present study. This arises from the outstanding resistance to hydrolysis of the oxide because of the presence of phosphates in the alumina structure [8].

In the case of thicker films, such as those produced by chromic acid anodising and sulphuric acid anodising, the corrosion-resistance properties of these oxides is a consequence of the manner in which the hydration process proceeds, and does not arise from the basic resistance of the oxide to hydrolysis. The important feature for the CAA and SAA produced oxides is that the thickness of these oxides is many times greater than their pore diameter, thus it is very difficult for a fresh supply of water to reach the base of the pores. Thus, at the mouth of the pore (at the surface of the oxide) where the supply of fresh water can be replenished easily then hydration occurs readily in this location. This leads to a "plug" being formed in the pore which essentially seals the pore. Now water cannot readily enter the pore and so this prevents hydration of the underlying oxide, resulting in the continued corrosion protection of the aluminium. However, it shoud be noted that the intrinsic hydration resistance of the SAA oxide is improved by the phosphoric-acid dip treatment, since an outer coating of insoluble aluminium phosphate is now formed by the SAA/PAD pretreatment. Additional data on the hydration characteristics of the thicker oxides has been published previously [9].

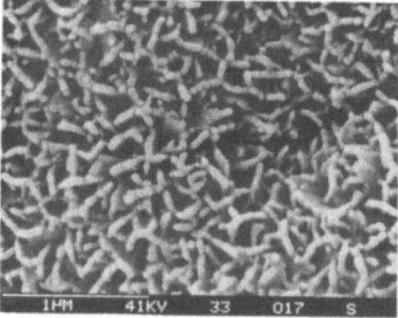

Figure 6   Scanning electron micrograph of aluminium surface, pretreated by CAE etch, after 24 hours exposure to water at $55^{\circ}C$ showing the formation of aluminium hydroxide platelets ('cornflakes') produced by the extensive hydration.

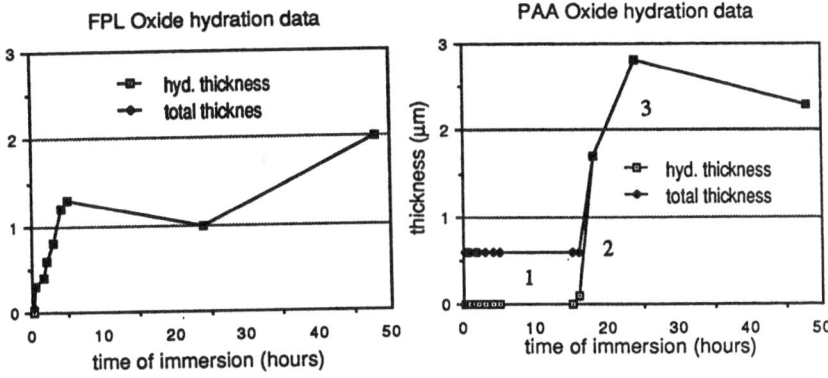

Key :   1. Resistance to hydration.
         2. Nucleation and growth of hydroxide.
         3. Decreasing rate of hydration.

Figure 7 Hydration results of (a) the CAE oxide and (b) the PAA oxide, as a plot of total thickness (thickness of any hydroxide present and any remaining oxide) and hydroxide thickness versus time of immersion in distilled water at 55°C

Characterisation of the Oxide/Primer Interface

The aim of this part of the study was to employ high-resolution scanning electron microscopy and transmission electron microscopy to determine the degree of penetration of the primer into the oxide to form a "micro-composite" interphase structure. The results from such studies indicates that substantial penetration of the open, porous oxide produced by the PAA treatment is possible (Figure 8) but far more limited penetration into the partly-sealed branch-like oxide produced by the CAA treatment occurs (Figure 9). In the case of the SAA treatment, then the oxide consists of very long and narrow pores and little sign of primer penetration is evident.

Now if specimens consisting of pretreated aluminium coated with a layer of oxide are immersed in water at 55°C for up to 4,000 hours then no indications of any gross hydration of the oxides under a polymeric primer are observed. This should be compared to the free-oxide hydration studies, discussed above, which revealed that these oxides readily hydrate when not covered by a polymeric layer. Hence, no hydration of the oxide occurs under a polymer, or if it does occur it is a very subtle process. Thus, one can conclude that any hydrate product observed, such as that shown in Figure 6, on the surface of failed aluminium-alloy joints is a result of post-failure hydration of the fracture surfaces, rather than being a direct cause of the environmental failure of the bonded joint as suggested in previous work [6].

16

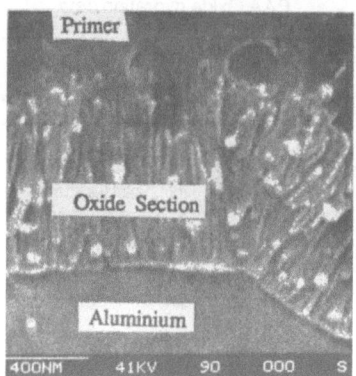
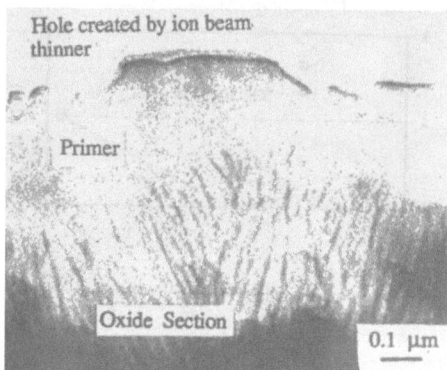

Figure 8 Cross section of the PAA oxide within a joint a) SEM micrograph and b) TEM micrograph. The SEM micrograph shows charging of polymer, and this occurs deep into the oxide pores. The TEM micrograph shows the open porous structure of the oxide and evidence of ion beam damage to the polymer within the oxide.

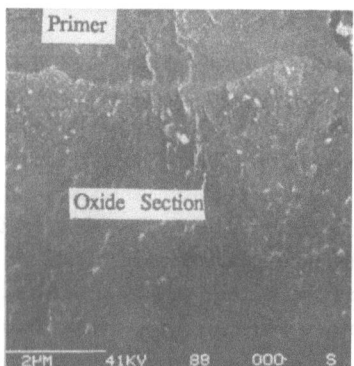
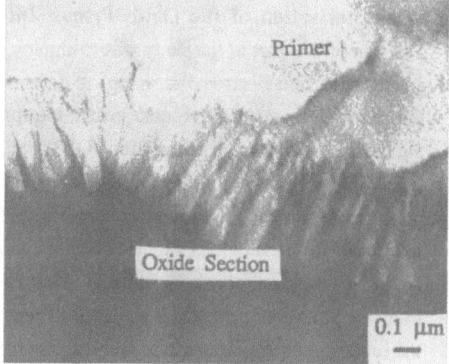

Figure 9 Cross section of the CAA oxide within a joint a) SEM micrograph and b)TEM micrograph.The SEM micrograph shows evidence of limited polymer penetration, as charging of the polymeric primer is only evident in the upper regions of the oxide thickness. The TEM micrograph shows the branch-like stucture of the oxide.

Durability Studies

To ascertain the durability of the bonded aluminium-alloy joints a novel accelerated test method has been devised. Essentially, double-cantilever beam (DCB) joints were prepared by bonding two beams of the pretreated and primed metal substrates and a pre-crack, 30mm in length, was inserted into the adhesive layer by loading the joint at a constant displacement rate using a tensile testing machine. The joint was then placed in a creep-rig apparatus which enabled the joint to be kept under a constant load. Further, the joint was surrounded by a tank of water and the water circulated through the tank at 55° C. The pre-cracked joint was loaded initially to 40 kg which was consequently increased by 5.6 kg in a stepwise fashion every 125 hours. (This method was adopted to overcome the problem of the crack in the adhesive layer blunting out, consequently reducing the local stress concentration and thus greatly delaying the eventual environmental failure of the joint). The time for failure of the joint to occur was monitored electronically.

The results for primed joints are shown in Figure 10 and are expressed in terms of the applied fracture energy, G_c(joint), versus the time to failure, t_f. The mechanism of crack growth and joint failure were also observed. Essentially, the initial cohesive pre-crack appeared to remain stationary and did not propagate. Instead, a new crack initiated at, or near, the primer/oxide interface. However, after growing a short distance this "interphase" crack flipped back into the adhesive layer and then propagated rapidly through the adhesive layer to give complete joint failure. Thus, the time to failure, t_f, is virtually completely associated with an incubation period for the new interphase crack to initiate and the time for this crack to grow along the interphase. The typical scatter on the measured time to failure for a given pretreatment was within $\pm 5\%$. Thus the newly devised accelerated test method seems to be able to determine the durability performance of the various pretreatments within strict limits. Also, and most importantly, it can discriminate between different pretreaments in a comparatively very short time period and only requires a small number of joint specimens.

Thus, the durability experiments clearly rank the order of joint durability as:

$$PAA \cong SAA/PAD > CAA > SAA > CAE.$$

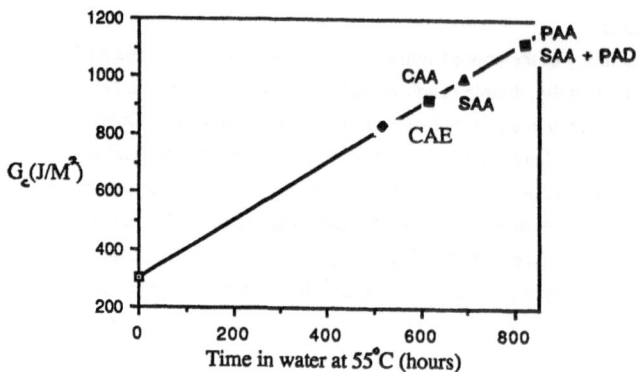

<u>Figure 10</u>   Durability data from DCB joints; longer the time to failure the more resistant is the joint to water attack; initial G_c applied of 300 J/m^2 is 95% of "dry" G_{Ic}.

Locus of Failure Studies

The locus of failure of the joints was examined using scanning electron microsopy and X-ray photoelectron spectroscopy. Another major advantage of using the fracture mechanics specimens is that the presence of a large crack in the adhesive layer results in the interface around the crack being subjected to a high stress concentration and hence suffering environmental attack preferentially. Thus, one knows that to determine the locus of failure due to the presence of the attacking environment the interfacial regions in the vicinity of the starter crack are those regions of primary interest. (In a joint such as a single-lap joint the area where environmental attack is likely to have initiated, and hence needs to be interrogated in detail, is not so readily discernable.)

The analyses of the fracture surfaces of the double-cantilever beam joints revealed that the environmental attack of the loaded joint by water clearly begun in the interphase regions above and below the starter crack that was placed in the adhesive layer. Small areas of apparently interfacial failure are seen just ahead of the initial crack front, the rest of the failure surface being cohesive. The actual locus of these critical areas of apparently interfacial failure, where environmental attack had occured, was determined by a combination of scanning electron microscopy and XPS. The results of these locus of failure studies for the different joints made using the different pretreatments are summarised in Table II.

TABLE II
Locus of failure for DCB Joints.

Treatment	Locus of Failure in DCB Joint
CAE	Oxide "pull out" from aluminium.
PAA	Through micro-composite primer/oxide interphase, just below surface of oxide.
CAA	Through micro-composite primer/oxide interphase, 0.5 to 1μm below surface of oxide.
SAA	Through hydroxide "crust" on top of original oxide.
SAA/PAD	Through oxide peaks created by post etch, so again through micro-composite primer/oxide interphase.

Mechanisms of Environmental Attack

In considering the mechanisms of attack that may be proceeding at the interphase in a joint then several observations may be made.

Firstly, the hydration of the oxide, as shown in Figure 6, appears to be a post-failure event rather than a mechanism of environmental attack. This is not to say that a more subtle transformation of alumina adjacent to any polymer may not take place. However, we have not detected such a more subtle process. The conclusion therefore must be that the importance of the hydration, and subsequent weakening, of the oxide as a main mechanism of environmental failure has yet to be conclusively established.

Secondly, the penetration of an adhesive into a oxide morphology creating a "micro-composite" interphase within the oxide region may have a profound influence on the durability performance of a joint for several reasons. This is not to invoke a simplistic idea such as mechanical keying as a primary mechanism of adhesion for these interfaces [1] but, for example, if the oxide can be substantially wetted and penetrated then any disruptive effect of the ingressing moisture would be delayed due to an increase in the diffusion path. Also, the actual stress states in the interphase region will depend on the mechanical properties of the adhesive, micro-composite interphase and oxide and this will obviously change as water ingresses into the region. Again the importance of these aspects has yet to be established.

Thirdly, no evidence has yet been found from the present work, and indeed from previous studies, to demonstrate the presence of any interfacial chemical bonds. Although the frequent observation that a phenolic primer may impart good joint durability has been ascribed to the formation of an interfacial chelate structure [1].

Thus, the authors would simply observe that the detailed mechanisms of environmental attack of water upon aluminium-alloy joints, and explanations for the different durability behaviour of joints prepared using different surface pretreatments, still alludes the adhesion scientist.

CONCLUSIONS

1.   The use of a novel test method for determining the durability performance of aluminium joints has been described. The new accelerated test method seems to be able to determine the durability performance of the various pretreatments with only a low degree of scatter in the experimental data. Also, and most importantly, it can discriminate between different pretreaments in a comparatively very short time period and only requires a small number of joint specimens.

2.   The ranking of the joint durability in water is clearly a function of the surface pretreatment used for the aluminium alloy and the order of performance is:

$$PAA \cong SAA/PAD > CAA > SAA > CAE$$

It should be emphasised that in this work we used a low-viscosity primer in preparing the joints, and the ranking would most likely be different if the primer was omitted.

3.   Phosphorus in the form of phosphates is present in oxides prepared in a phosphoric acid solution but no chromium is present in oxides formed in a chromic acid solution. The role of such ions in the hydration resistance of the free oxide structure, together with the effects of oxide thickness, has been described.

4.   However, hydration of the oxide, such as that seen on the free oxide (figure 6), has not been observed and appears therefore not to be involved in the environmental failure mechanism. Any such hydration observed on failed surfaces of joints that have fractured in the presence of water is a result o the post-failure exposure of the oxide to water.

5.   The locus of environmental failure for the DCB joints under investigation does appear to involve fracture through the oxide layer. But this is really better thought of as fracture through

the "micro-composite" interphase, and so usually involves the fracture of the penetrated primer as well as the oxide.

6. Various mechanisms of environmental attack have been considered but it is suggested that there is currently insufficient proof for identifying the details of the mechanisms responsible.

ACKNOWLEDGEMENTS

The authors wish to acknowledge the advice and assistance given by Mrs. S. Cummins, DQA/TS, MoD (PE), Woolwich, England on the XPS work and Mr. G. Seyd, ARE (HH), MoD (PE), Poole, England on the XRD work. Also the authors gratefully acknowledge the financial support given by the U.K Ministry of Defence (Procurement Executive).

REFERENCES

1. Kinloch, A. J., "Adhesion and Adhesives: Science and Technology", Chapman and Hall, London, 1987.

2. Kinloch, A. J., "Durability of Structural Adhesives", Applied Science Publ., London, 1983.

3. Poole, P., "Industrial Adhesion Problems", Ed. D.M. Brewis and D. Briggs, Orbital Press, Oxford, 1985, p.258.

4. Arrowsmith, D. J., Clifford, A. W., Moth, D. A. and Davies, R. J., "The Adhesive Bonding of Aluminium-Lithium Alloys", Proc. 3rd Al-Li Conf., University of Oxford, July 1985.

5. Vedder, W. and Vermilyea, "Aluminium and Water Reaction", Trans. Faraday Soc., (1969), 65, 563.

6. Davis, G. D. and Venables, J. D., "Durability of Structural Adhesives", ed. Kinloch, A. J., Applied Science Publ., London, 1983, page 43.

7. Hunter, M S, Towner, P F and Robinson, D L, "Hydration of Anodic Oxide Films", Proc.Amer. Electroplaters Soc., (1959), **46** , 220.

8. Xu, Y., Thompson, G. E. and Wood, G. C., " Mechanisms of Anodic Film Growth on Aluminium", Trans. Inst. Metal Finishing, (1985), **63**, 98-103.

9. Davies, R J, "The Morphology and Properties of Aluminium Oxides", Proc.Int. Adhesion Conf., York University, U.K, Sept. 1987.

3

SURFACE FREE ENERGY CALCULATION

FOR POLYMERS BY A GROUP CONTRIBUTION METHOD

A. Carre and J. Vital

Centre Européen de Recherche et Technologie

Corning Glass

7bis, avenue de Valvins, 77210 Avon,

FRANCE

1. INTRODUCTION

The surface free energy, γ, of a polymer is an important parameter for understanding, interpreting and predicting numerous surface phenomena such as adsorption, wetting, adhesion, sliding friction....

Poor mobility of molecules in a polymer does not allow one to determine directly its surface energy. Such determination currently implies indirect methods involving wettability measurements [1-7]. These methods are based on numerous hypotheses allowing expression of the contact angle of a liquid as a function of the surface energies of the liquid and of the solid.

23

In this study, a different approach is proposed. It consists of considering the surface energy of a series of liquids having the same chemical structure as the polymer. These particular liquids will be called "homostructural liquids". The principle of the calculation of γ amounts to considering the contribution of each chemical group to the surface energy of the material.

The basis of the calculation will be described with a few examples and results will be compared with the values of surface energy determined by means of contact angle measurements.

2. PRINCIPLE

It will be assumed that surface energy results from the sum of contributions from each chemical group component of a molecule. Calculation of solubility parameter has been successfully developed by Small [8] who used a similar group contribution method.
Let us consider a liquid L having the following chemical pattern:

$$A-(B)_{n-2} - A \qquad\qquad L$$

We want to express the surface energy of L, γ_n as a function of n and of the group contributions, γ_A and γ_B. If we suppose L to be a mixture of A and B groups, and if we admit a statistical distribution of A and B at the surface, the number of A groups at the surface is proportional to 2/n and B to n-2/n.

As a consequence, the surface energy of L will be expressed
by an arithmetic mean relationship with the form:

$$\gamma_n = \frac{1}{n}\left[2\gamma_A + (n-2)\gamma_B\right] \qquad 1$$

The accuracy of this simple expression will be verified by
using several series of homostructural liquids.

With the same hypothesis, equation 1 can be generalized for a
multi-group compound. If A, B, C.... are these groups, the
surface energy can be written:

$$\gamma_n = \frac{1}{n}\left[a\gamma_A + b\gamma_B + c\gamma_C +..\right] \qquad 2$$

a, b, c... being the number of groups A, B, C... forming the
molecule, ie. a + b + c ... = n.

Calcuation of the Surface Free Energy of a Polymer

In a macromolecular material, n is very large, so that the
contribution of the end-group of chains to the surface energy
tends to be negligible. For a polymer homostructural of L,
if n is considered to be infinite, equation 1 becomes:

$$\gamma_\infty = \gamma_B \qquad 3$$

This leads us to write for any value of n:

$$\gamma_n = \frac{1}{n}\left[2\gamma_A + (n-2)\gamma_\infty\right] \qquad 4$$

If we study the variation of γ_n for a series of homostructural
liquids, equation 4 implies than $n\gamma_n$ must be a linear

function of n-2, the slope of the line being precisely the surface energy, γ_∞ of the polymer homostructural to the liquid series considered.

In the following examples, relation 4 will be explicited so that each term will have a physical meaning, ie. the lowest value of n will correspond to the first member of the homostructural series existing as a liquid at the temperature considered.

Several homostructural liquid series have been examined, so that the surface energies of Polythylene (PE), Polytetrafluoroethylene (PTFE), Polydimethylsiloxane (PDMS) and Polypropylene (PP) have been calculated. In addition, the variation of the surface energy of PE with temperature is presented.

Application

The method described required the knowledge of the surface energies of several pure liquids homostructural to the polymer. For example, with PE, the simplest series of liquids is that of the normal alkanes of structure:

$$H_3C - (CH_2)_{n-2} - CH_3 \qquad\qquad I$$

n being the number of carbon atoms in the molecule.

By using the generalised equation 2, the calcuation of γ_{PE} will be also examined by considering the chain length dependence of γ_n for n-aliphatic primary alcohols (II) and for 1-nitroparaffins (III). These series displaying slightly polar character are of particular interest.

The surface energy of PTFE will be estimated from the surface energy of perfluorinated alkanes:

$$F_3 C - (CF_2)_{n-2} - CF_3 \qquad\qquad IV$$

PDMS and PP display more complex structures. But by using the equation 2 and surface tension of approximately homostructure liquids, the surface energy of these polymers can nevertheless be calculated simply.

The surface energy of a homostructural liquid is easily measured by tensiometry and the values are cited in the literature. Jasper [9] has made a list of over two thousand pure liquids. Tables A-1 to A-6 in the Appendix give the surface energies of the liquids useful in the calculation of γ_{PE}, γ_{PTFE}, γ_{PDMS}, and γ_{PP}.

1.   Polyethylene

1.a   Calculation of γ_{PE} from a series of non polar liquids

Surface energies of n-alkanes at several temperatures are summarized in Table A-1 for n between 5 and 36. At 20°C the compounds are not liquid if n <5. In order to give a physical meaning to the members of equation 4, γ_n has to be expressed first for n=5:

$$\gamma_5 = \frac{1}{5}\left[2\gamma_{CH3} + 3\gamma_\infty \right] \qquad\qquad 5$$

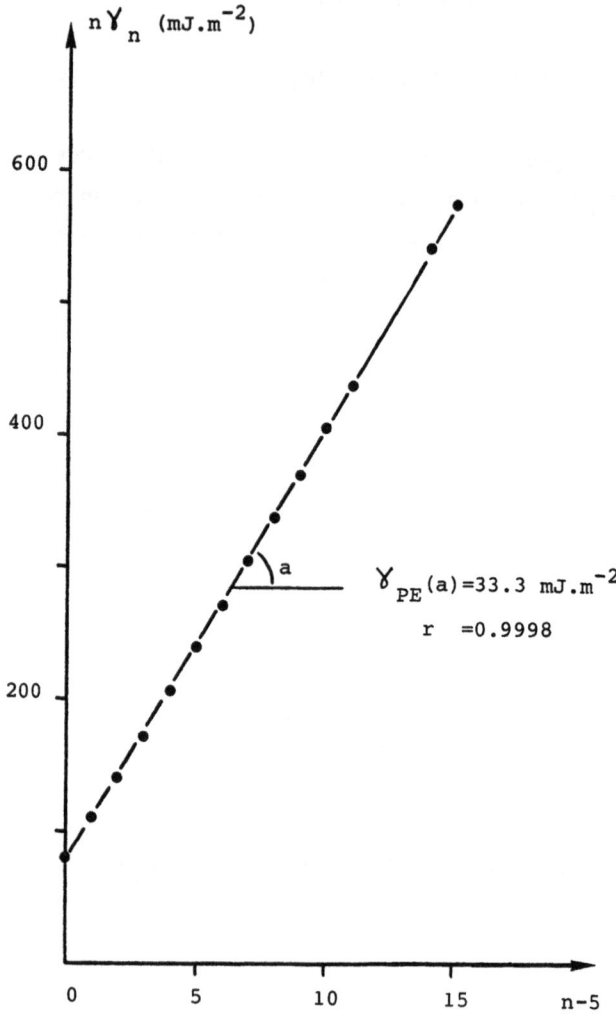

Figure 1. Linear relationship between $n\gamma_n$ and n-5 for n-alkanes. Determination of surface energy of PE at 20°C.

Thus, for n >5, we should verify that:

$$\gamma_n = \frac{1}{n} \left[5\gamma_5 + (n-5)\gamma_\infty \right] \qquad \qquad 6$$

OR $\qquad \qquad n\gamma_n = 5\gamma_5 + (n-5)\gamma_\infty$

Figure 1 shows the variation of $n\gamma_n$ at 20°C as a function of n-5, for n varying from 5 to 20. A very good linear relationship is obtained with a correlation coefficient of the linear regression, r, of 0.9998. The slope, a, representing γ_∞, will be taken as the surface energy of the PE, ie. 33.3 mJ.m^{-2} at 20°C.

1.b Calcuation of γ_{PE} from a series of polar liquids

γ PE has also been tentatively determined from n-aliphatic primary alcohols, $CH_3-(CH_2)_{n-2}-OH$ (II), and 1-nitroparaffins, $CH_3-(CH_2)_{n-2}-NO_2$ (III). Equation 2 leads to the following relationship for liquids II and III:

$$\gamma_n = \frac{1}{n} \left[2\gamma_2 + (n-2)\gamma_\infty \right] \qquad \qquad 7$$

A graphical representation of equation 7 is given in Figures 2 and 3 ($n\gamma_n$ versus n-2). It appears that, contrary to n-alkanes, a well correlated linear relationship is obtained only above a given chain length. A summary of the data, minimum values of n, γ_∞ and r, is given in Table I (γ_n values are in Tables A-2 and A-3). Under these conditions, γ_∞ values do not differ very much from the value determined with n-alkanes (less than 13%).

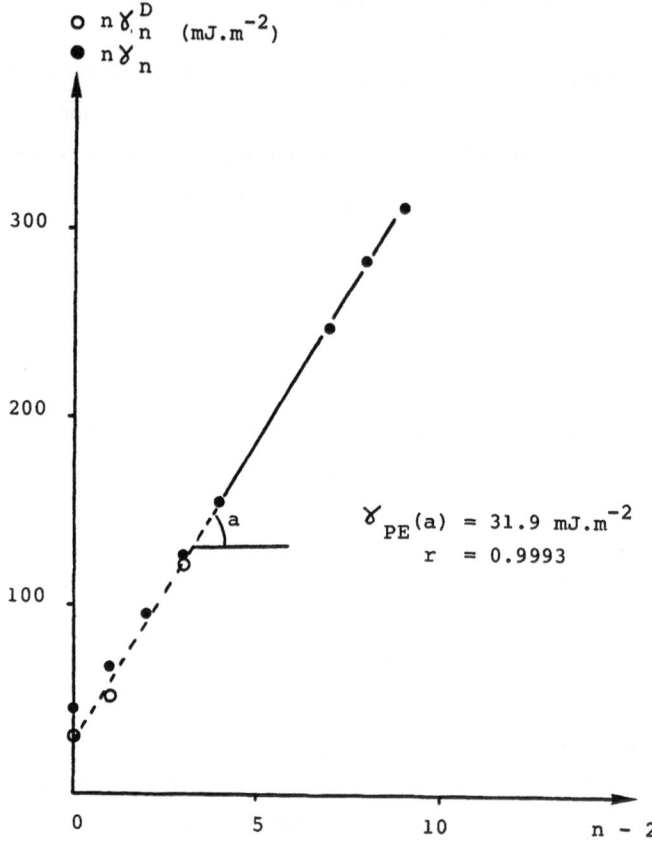

Figure 2.   Linear relationship between $n\gamma_n$, $n\gamma_n^D$ and n-2
            for n-aliphatic primary alcohols.

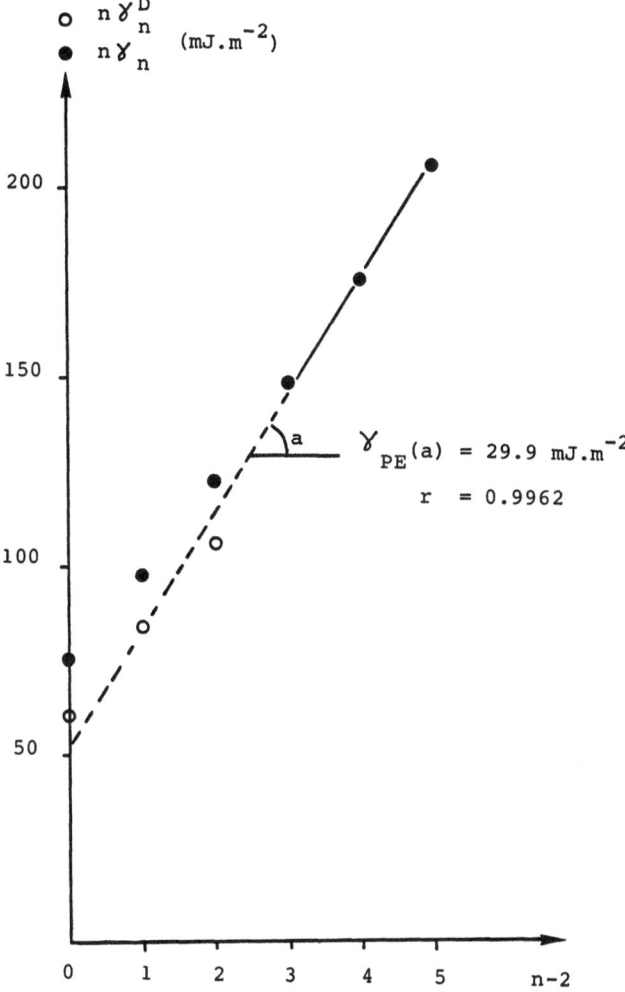

Figure 3. Linear relationship between $n\gamma_n$, $n\gamma_n^D$ and n-2 for 1-nitroparaffins.

TABLE 1

Range of Linearity of $n\gamma_n$ versus n-2 for n-aliphatic primary alcohols and 1-nitroparaffins

Liquids	n	$\gamma\infty$ (mJ.m^{-2})	r
n-aliphatic primary alcohols	>5	31.7	0.9998
1-nitroparaffins	>4	29	0.9992

The deviation from the straight line for the liquids having the shorter chain lengths can be explained if we consider that the molecules of structure II and III display a non-negligible polarity. Intemolecular forces are both dispersive and polar in nature. N-aliphatic primary alcohols interact also b hydrogen bonding. This is obviously not the case with n-alkanes for which the surface energy results only from dispersive interactions.

For polar or hydrogen bonded liquids of type II and III, the dispersive character increases with chain length. The values of the ratio $\gamma^D/$, (γ^D being the dispersive component of γ) are given in the Tables A-2 and A-3. For a totally dispersive liquid, γ^D/γ is equal to 1. Thus, high molecular weight compounds of structure II and III tend to have the same chain length dependence as aliphatic alkanes. This conclusion is supported by the fact that a fairly good correlation following equation 7 is gained when the values of the shorter molecules are substituted by the dispersive components of γ_n, γ^D_n. In figures 2 and 3 n γ^D_n instead of n γ_n has been plotted against n-2 for the lowest molecular weight compounds. $\gamma\infty$ is then calculated with very good agreement (r>0.996). Series II and III lead to $\gamma\infty$ respectively equal 31.9 and 29.9 mJ.m^{-2} for Polyethylene.

As a consequence, it can be taken that non-dispersive interaction in a liquid cannot be simply related to the number of polar groups. The distance between polar interacting groups has to be taken into account and the very simple model for the description of γ_n, based on an average calculation, is then not valid. This is less critical for the dispersive interactions which are always present whatever the nature of the interacting groups.

1.c.   Temperature dependence of γ_{PE}

The temperature dependence of the surface energy of PE has been calculated from the values of γ_n for n-alkanes at different temperatures from 0 to 120°C (Table A-1). Figure 4 gives three examples of the linear relationship 7 for T equal to 20°, 70° and 120°C. The values of γ_{PE} between 0° and 120°C are presented in Table II.

The calculation allows us to verify that γ_{PE} decreases when T increases and γ_{PE} is obtained with a good correlation coefficient in the overall temperature range.

In addition, the derivation of dγ_{PE}/dT can be estimated by

plotting γ_{PE} as a function of T (Figure 5). The slope of the straight line leads to dγ_{PE}/dT = -0.060 mJ.m^{-2}.K^{-1}.

TABLE II : Surface energy determination for Polyethylene from 0 to 120°C (mJ.m^{-2}) with n-alkane series

T	0	10	20	30	40	50	60	70	80	90	100	110	120°C
γ_{s}	34.8	33.2	33.3	32.6	32	31.3	30.5	30.3	29.6	28.9	28.4	27.9	27.4
r	0.9999	0.9986	0.9998	0.9998	0.9999	0.9999	0.9999	0.9999	0.9999	0.9999	0.9999	0.9998	0.9997

$$\frac{d\gamma_{s}}{dT} = -0,060 \text{ mJ.m}^{-2}.\text{K}^{-1}$$

(r = - 0.9948)

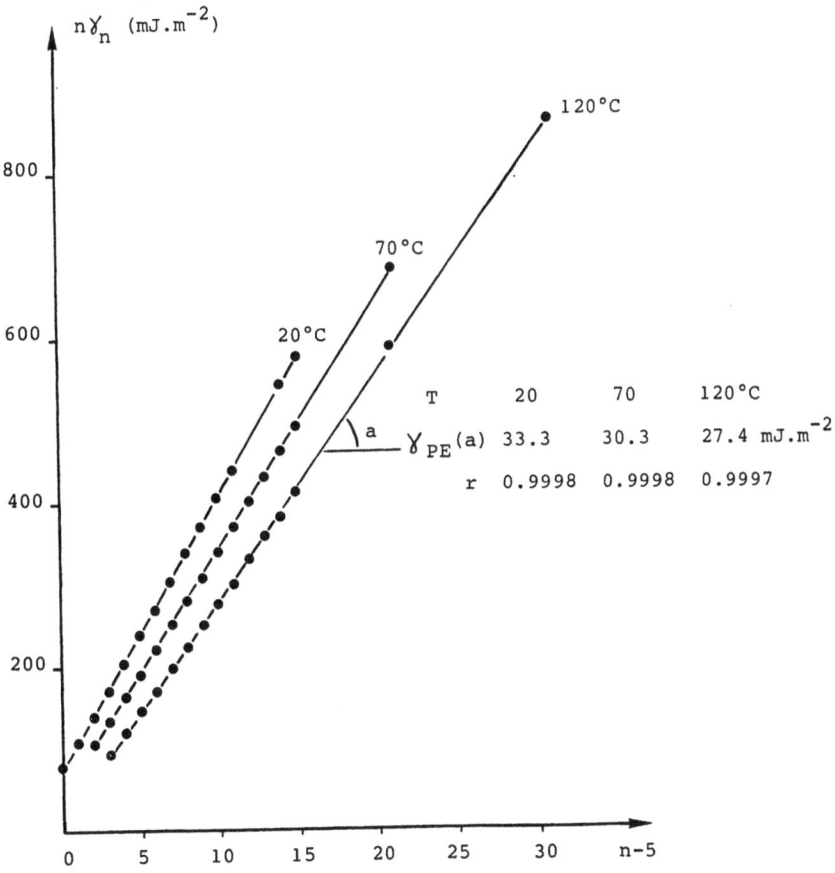

Figure 4. Linear relationship between $n\gamma_n$ and n-5 for n-alkanes and calculation of γ_{PE} at three different temperatures.

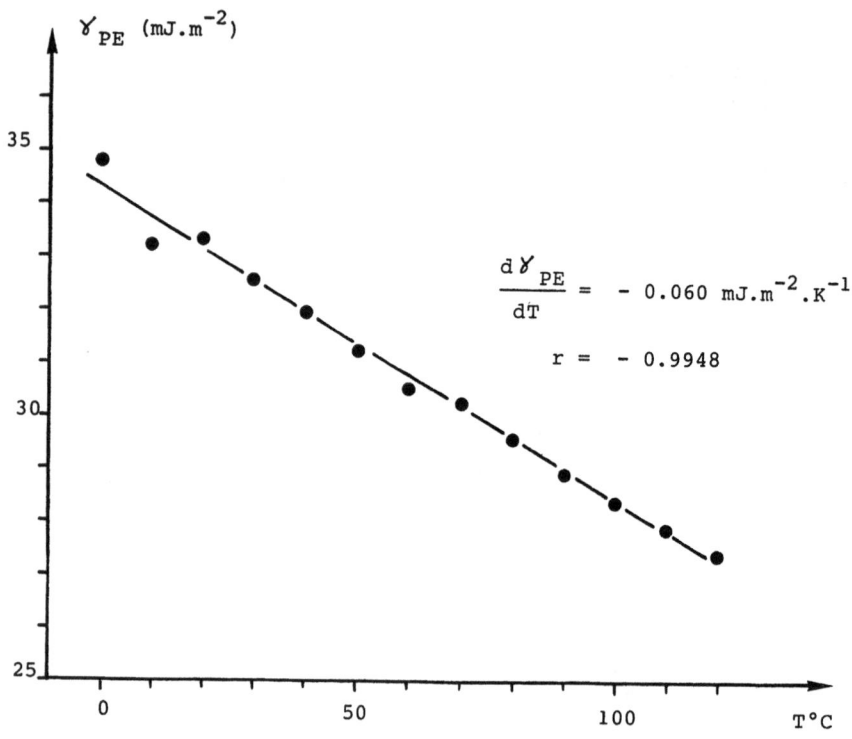

Fig. 5: $d\gamma/dT$ determination for PE

This value has been found in good agreement with the result of the derivation of the Macleod relationship leading to d $_{PE}$/dT equal to -0.069 $mJ.m^{-2}.K^{-1}$ at 100°C (details of the calculation are given in reference 10).

2. Polytetrafluoroethylene

At 25°C, five values of γ_n are available for perfluoro alkyl compounds of formula $C_n F_{2n+2}$ (IV) with n = 5,6,7,8 and 9. These values are reported in Table A-4. Some other values of γ_n are given for 15°, 20° and 45°C. As in 1.a the calculation of γ_n using γ_{CF_3} and γ_{CF_2} contributions leads to the following relation:

$$\gamma_n = \frac{1}{n}\left[5\gamma_5 + (n-5)\gamma_\infty \right] \left.\right\} 8$$

which becomes: $n\gamma_n = 5\gamma_5 + (n-5)\gamma_\infty$

with v>5 and $\gamma_\infty = \gamma_{PTFE}$

The variation of $n\gamma_n$ as a function of n-5 is plotted in Figure 6. The linear relationships are well defined with correlation coefficients very close to 1 and slopes equal to 21.7, 21.4, 20.6 and 19.3 $mJ.m^{-2}$ corresponding to the surface energy of PTFE at 15°, 20°, 25° and 40°C.

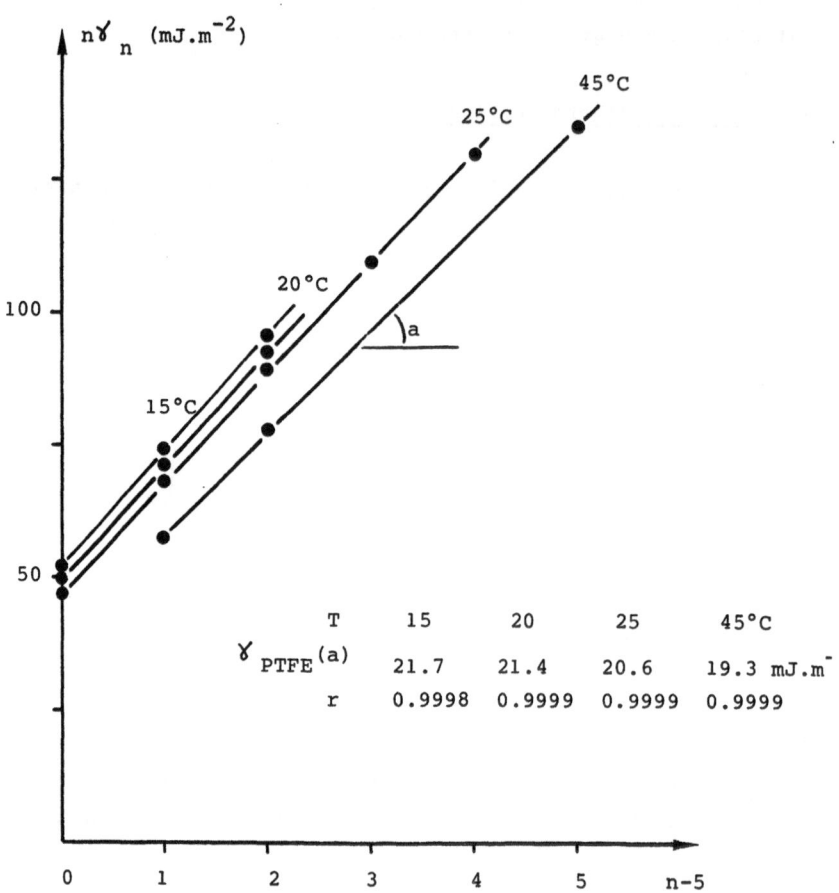

Figure 6. Linear relationship between $n\gamma_n$ and n-5 for perfluorinated alkanes. Determination of surface energy of PTFE at 15, 20, 25 and 45°C.

3. Polydimethylsiloxane

The liquid polydimethylsiloxanes considered present four different groups in the structural pattern:

$$
H_3C - \left[\begin{array}{c} CH_3 \\ | \\ Si - O \\ | \\ CH_3 \end{array} \right]_m - \begin{array}{c} CH_3 \\ | \\ Si - CH_3 \\ | \\ CH_3 \end{array} \qquad V
$$

We assume that the general relationship 2 is valid in the present case and we obtain:

$$
\gamma_m = \frac{1}{4m+5} \left[9\gamma_1 + 4\,(n-1)\gamma_\infty \right] \qquad 9
$$

The values for γ_m with n equal to 3, 4, 5, 6, 7, 9, 12 and 17 are reported in Table A-5. The relationship 9 can be transformed in the following way:

$$
\gamma_n = \frac{1}{4m+5} \left[17\gamma_3 + 4\,(\,.-3)\gamma_\infty \right] \qquad 10
$$

Figure 7 shows that equation 10 is well verified (correlation coefficient equal to 0.9999), when the surface energy of PDMS is expressed as a linear contribution of each constituent group ($-CH_3$, $-Si-$, $-O-$) of the molecules.

The calculated value of γ_{PDMS} is 20.8 $mJ.m^{-2}$.

40

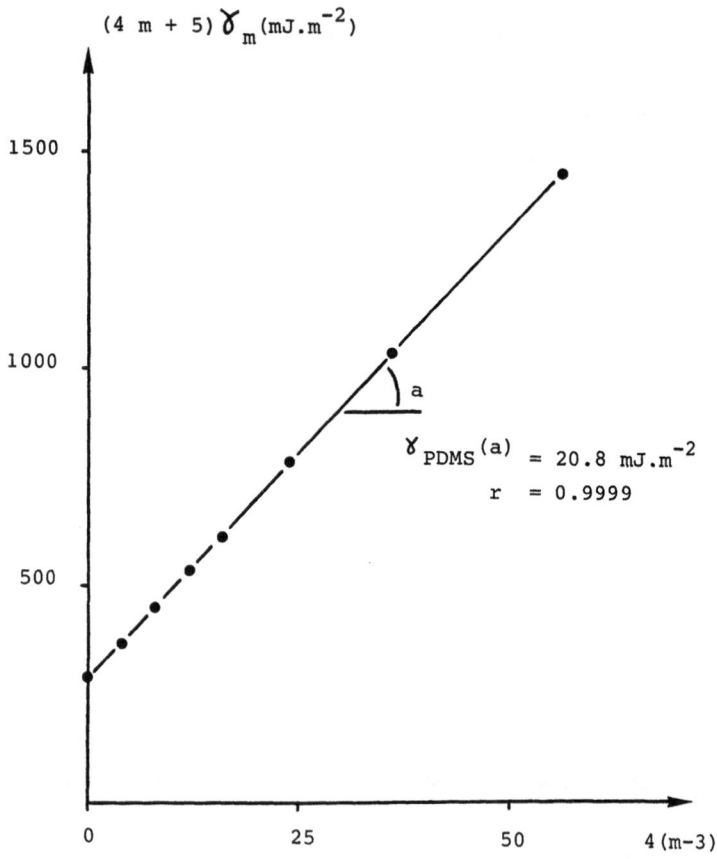

<u>Figure 7</u>. Linear relationship for polymethylsiloxanes.
Determination of surface energy for PDMS at 20°C.

4. Polypropylene

Homostructural liquids corresponding to PP have the general
following formula:

$$H_3C - \left[CH_2 - \underset{\underset{CH_3}{|}}{CH} \right]_p - CH_3 \qquad\qquad VI$$

As in the previous cases, the general equation 2 leads to the
following equation for γ_p:

$$p = \frac{1}{3p+2} \left[2\gamma_{CH_3} + 3_p\gamma_\infty \right] \qquad 11$$

γ_p values for p = 1 and 2 are obtained from Jasper[9] (Table
A-6).

If p = 1, the relationship 11 becomes:

$$\gamma_1 = \frac{1}{5} \left[2\gamma_{CH_3} + 3\gamma_\infty \right] \qquad 12$$

For the range referring to actual liquids at ambient
temperature (p >1), equation 11 expressed as a function of
γ_1 becomes:

$$\gamma_p = \frac{1}{3p+2} \left[5\gamma_1 + (3_p-3)\gamma_\infty \right] \qquad 13$$

The mathematical resolution of the system of two equations
corresponding to p = 1 and p = 2 gives γ_{pp} = 28.5 mJ.m^{-2}.

3. DISCUSSION

If polymers are considered as liquids of infinte molecular weight, the group contribution method does not take into account the crystallinity of polymers and its effects on surface properties. However, three of the polymers considered are particularly crystalline, depending on their thermal treatment, and are scarcely amorphous. The calculated values of surface energy correspond to ideal and hypothetic materials, but the real polymers always present structural defects, polydispersity and sometimes chain orientation at the surface.

However, we assume, to a first approximation, that crystallinity structural and composition defects have only a secondary effect on the statistical surface distribution of the molecular groups, and consequently negligible effect on surface properties.

Our method gives calculated values very close to those obtained by the experimental wetting methods as indicated in Table III. Our study confirms a well known result, that PTFE and PDMS are very low surface energy materials. Such values explain the anti-adhesive properties of these products when used as coatings.

4. CONCLUSION

From this orginal and simple approach, a method to calculate the surface energy of polymers has been developed using molecular group contributions of the macromolecular patterns. A linear relation has been established between the surface energy and the length of homostructural molecules and verifications made in several cases.

From this method it ispossible to calculate the theoretical surface energy of Polythylene, Polypropylene,

TABLE III : Experimental surface energy (from contact angle measurements) and theoretically calculated values for polymers at 20°C ($mJ \cdot m^{-2}$).

Polymer	γ_S	Method of Zisman (1,2) a	Method of Fowkes (4)	Method of Kaelble (5)	Method of Schultz & Coll (6) b	
Polyethylene	33.3[1], 31.9[2], 29.3[3]	31	35	32.4	34.5	31
Polytetrafluoroethylene	21.4	18.5	19.5	15.5	19	20
Polydimethylsiloxane	20.8	24	-	-	-	-
Polypropylene	28.5	29	-	-	-	-

1 from n-alkane,   2 from n-aliphatic primary alcohol,   3 from 1-nitroparaffin series

a : γ_C values. For a non-polar solid $\gamma_C \simeq \gamma_S$

b :   The two series of values correspond to two different experimental methods.

Polytetrafluoroethylene and Polydimethylsiloxane. The
experimental methods using wetting give very close results.

For Polythylene, the surface energy variation as a function
of temperature, has been determined. It is in good agreement
with values obtained from the Macleod relation.

ACKNOWLEDGEMENTS

The authors thank Dr M.E.R. Shanahan for helpful discussions
and for his help in the English version.

REFERENCES

1. Zisman, W. A. "Contact Angle, Wettability and Adhesion",
    Advances in Chemistry Series, Am. Chem. Soc. Washington
    DC, 1, 1964.

2. Shafrin, E. G. "Polymer Handbook", 2nd Ed., ed.
    Brandrup, J. and Immergut, E. H. John Wiley and Sons,
    Inc. NY., 1975.

3. Good, R. J. ibid (1), 74.

4. Fowkes, F. M. Ind. Eng. Chem., 56, 40, 1964.

5. Kaelble, D. H. "Physical Chemistry of Adhesion", Wiley
    Interscience, NY., 1971.

6. Carre, A., Schultz, J., Simon, H., 5° Conf. Europeene des
    Plastiques et des Caoutchoucs, Paris, 1, C-17, 1978.

7. Schultz, J., Tsutsumi, K., Donnet, J. B., J. Colloid
    Interface Sci., 59, 272, 1977 - 59, 277, 1977.

8. Small, P. A. J. Appl. Chem., 3, 71, 1953.

9. Jasper, J. J. J. Phys. Chem., Ref. Data 4, 841, 1972.

10. Carre, A., Vial, J. J. Chim. Phys., To be Published 1988.

11. Carre, A., These de Doctorat d'Etat, Mai 1980.

Surface energy of liquids (9)

Table A-1 : Surface energy of n-alkanes at different temperatures ($\pm$ 0.10 mJ.m^2)

Alkane	10°C	20°C	30°C	40°C	50°C	60°C	70°C	80°C	90°C	100°C	110°C	120°C
Pentane ------	17.15	16.05	14.94	---	---	---	---	---	---	---	---	---
Hexane ------	19.42	18.40	17.38	16.36	15.34	14.32	---	---	---	---	---	---
Heptane ------	21.12	20.14	19.17	18.18	17.20	16.22	15.24	14.26	13.28	---	---	---
Octane ------	22.57	21.62	20.67	19.71	18.77	17.81	16.86	15.91	14.96	14.01	13.06	12.11
Nonane ------	23.79	22.85	21.92	20.98	20.05	19.12	18.18	17.24	16.31	15.37	14.44	13.50
Decane ------	24.75	23.83	22.91	21.99	21.07	20.15	19.23	18.31	17.39	16.47	15.55	14.63
Undecane ------	25.56	24.66	23.76	22.86	21.96	21.05	20.15	19.25	18.35	17.45	16.55	15.65
Dodecane ------	26.24	25.35	24.47	23.58	22.70	21.81	20.93	20.05	19.16	18.28	17.39	16.51
Tridecane ------	26.86	25.99	25.11	24.24	23.37	22.50	21.63	20.75	19.88	19.01	18.14	17.27
Tetradecane ---	27.43	26.56	25.69	24.82	23.96	23.09	22.22	21.35	20.48	19.61	18.74	17.87
Pentadecane ---	---	27.07	26.21	25.35	24.50	23.64	22.78	21.93	21.07	20.21	19.36	18.50
Hexadecane ---	---	27.47	26.62	25.76	24.91	24.06	23.20	22.35	21.49	20.64	19.79	18.93
Heptadecane ---	---	---	27.06	26.22	25.38	24.52	23.68	22.83	21.99	21.14	20.29	19.45
Octadecane ---	---	---	27.45	26.61	25.77	24.92	24.08	23.24	22.39	21.55	20.71	19.87
Nonadecane ---	---	28.59	27.75	26.91	26.07	25.24	24.40	23.56	22.73	21.89	21.05	20.21
Eicosane ------	---	28.87	28.04	27.21	26.38	25.54	24.71	23.88	23.04	22.21	21.38	20.54
Hexacosane ---	---	---	---	---	---	---	26.33	25.59	24.86	24.13	23.39	22.66
Hexacontane ---	---	---	---	---	---	---	---	---	---	---	24.48	23.90

APPENDIX (continued)

TABLE A-2 :   Surface energy of n-aliphatic primary alcohols
(mJ.m^{-2}) at 20°C and γ^D/γ ratio.

Alcohol	γ	γ^D/γ (*)
Methanol	22.50	0.68
Ethanol	22.39	0.74
Propanol	23.71	
Butanol	25.39	0.$\overline{9}$9
Pentanol	25.79	-
Octanol	27.50	-
Nonanol	28.27	-
Decanol	28.88	-

(*) : For the calculation of γ^D/γ see reference 11.

TABLE A-3 : Surface energy of 1-nitroparaffins (mJ.m^{-2})
at 20°C and γ^D/γ ratio.

Nitroparaffin	γ	γ^D/γ (*)
Nitromethane	37.48	0.81
Nitroethane	32.66	0.86
Nitropropane	30.64	0.87
Nitrobutane	29.77	-
Nitropentane	29.31	-
Nitrohexane	29.54	-

(*) : see reference 7

APPENDIX (continued)

TABLE A-4 : Surface energy of perfluorinated alkanes at different temperatures (mJ.m^{-2}).

Perfluoroalkane	15°C	20°C	25°C	45°C
Perfluoropentane	10.35	9.89	9.42	
Perfluorohexane	12.38	11.91	11.44	9.57
Perfluoroheptane	13.60	13.19	12.78	11.14
Perfluorooctane			13.70	
Perfluorononane			14.40	
Perfluorodecane				13.5

TABLE A-5 : Surface energy of polymethylsiloxane at 20°C (mJ.m^{-2})

Polymethylsiloxane

Trimere	17
Tetramere	17.6
Pentamere	18.1
Hexamere	18.5
Heptamere	18.5
Nonamere	19.2
Dodecamere	19.5
Heptadecamere	19.9

TABLE A-6 : Surface energy of 2-methylbutane and 2,4 dimethylhexane at 20°C (mJ.m^{-1})

Alcane

2-methylbutane	14.99
2,4-dimethylhexane	20.05

4

ROLE OF MOLECULAR DISSIPATION IN
ELASTOMER ADHESION

M.E.R. Shanaha n,[*] P. Schreck, J. Schultz
Centre de Recherches sur la Physico-Chimie des Surfaces Solides

and

Ecole Nationale Supérieure de Chimie de Mulhouse,
3, rue Alfred Werner,
68093 Nulhouse Cedex,
FRANCE

1. INTRODUCTION

Consider two optically flat solids put into intimate (molecular) contact. In the absence of interfacial chemical bonding, the forces of adhesion between them should be limited to those of a physical, or van der Waals, nature and, as a consequence, we would expect to be able to calculate the reversible work of adhesion, W_o, from Dupré's equation:

$$W_o = \gamma_1 + \gamma_2 - \gamma_{12} \qquad \cdot \quad \cdot \quad \cdot \qquad 1$$

where γ_1 and γ_2 are the free surface energies of the materials in contact and γ_{12} represents their common interfacial free energy.

Let us take the case where one of the materials is elastomeric and, for the sake of argument, the other may be regarded as infinitely rigid. Separation at the interface may be effected by peeling off the elastomer layer. If each material takes the form of a rectanguloid block the work of

[*] M.E.R. Shanahan present address:

Centre des Matériaux,
B P 87
91003 Evry Cédex,
FRANCE

adhesion, W, may be assessed from the classic equation:

$$W = F \left(1 - \cos \alpha \right) \qquad \cdot \quad \cdot \quad \cdot \qquad 2$$

where F is the peel force per unit width of peel front and α is the peel angle. It is generally found that the measured value of W is far greater than W_0 - typically by several orders of magnitude - even when interfacial bonding is only of a physical nature. The explanation at least partially resides in the fact that although interfacial separation involves the transfer of energy W_0, in order to provoke this separation, the elastomeric phase must be mechanically deformed. Linked to this deformation is an energy dissipation in the bulk of the polymeric phase, which generally far exceeds W_0.

Nevertheless, as W_0, increases, so does the deformation and thus the dissipation. Therefore, although $W \gg W_0$, there does exist a proportionality between them. The factor relating them is a function of both strain rate and temperature [2-4] and involves the well known principle of time-temperature superposition developed by Williams, Landel and Ferry [5] (WLF). This dissipation process explains matters at finite separation rates. Nevertheless, as the peel rate tends towards zero, so apparently should the hysteresis losses and one would expect the measured adhesion, W, to approach the thermodynamic value, W_0. However, peel experiments at extremely low rates of separation suggest that the threshold value of W is still far greater than W_0. It would therefore seem that a further type of dissipation is present on a molecular scale even when the bulk effect is absent.

Since it has been postulated that such a phenomenon could be connected with the average inter-crosslink molecular weight, M_c, of the elastomer [6-8], it was decided to investigate elastomer adhesion as a function of degree of cross-linking.

2. EXPERIMENTAL

2.1 Method

Most tests used for measuring adhesion, such as the classic peel test, are dynamic and as a consequence, a direct estimation of resistance to separation at zero speed is inaccessible experimentally. As a consequence, for the present study it was decided to investigate adhesion properties using the quasi-static technique originally developed by Johnson, Kendall and Roberts [9] (JKR).

When two spheres of an elastic nature are forced into contact, their common contact area may be calculated by the classic theory of Hertz dating back to the nineteenth century [10]. However, this early work totally ignored effects due to surface attraction, or adhesion, and it was not until the JKR analysis was developed that surface effects were taken into account. For spheres forced together under considerable pressures, these phenomena are of little importance, but when low loads are imposed, their effects can become predominant. If one sphere is considered to be of infinite radius, the problem becomes that of the contact of a sphere and a plane. This is the case adopted here in which the experimental technique consists of placing a small, flat, essentially rigid material in contact with the surface of an elastomeric hemisphere, as shown in Fig. 1. At equilibrium, the elastomer becomes slightly flattened and a circular contact area of radius a results. The value of a can be readily assessed using a low power microscope. Contact area is determined by the minimum of free energy of the system hemisphere/flat where the various contributions to the overall free energy stem from adhesion (W), elastic strain energy within the deformed hemisphere, and potential (gravitational) energy due to the weight of the flat (P).

The JKR analysis gives an expression for the work of adhesion, W:

$$W = \frac{3}{32\,\pi\,E\,a^3}\left(\frac{16\,E\,a^3}{9R} - P\right)^2 \qquad \cdot \quad \cdot \quad \cdot \qquad 3$$

where E is the Young's modulus of the elastomer and R is the hemispherical radius. Apart from direct measurement of a, R and P, we also need a value of E, determined as described below.

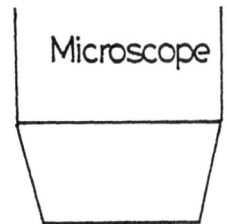

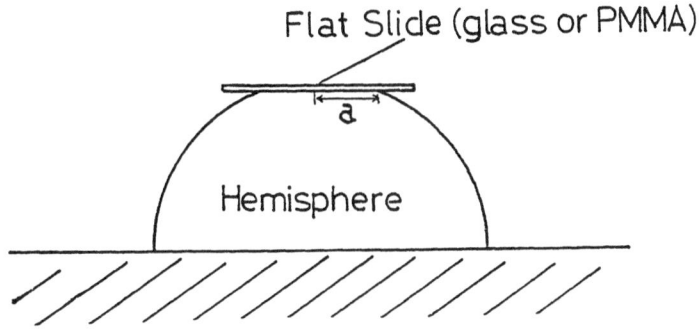

Figure 1. Experimental technique in which contact area radius, a, of flat
slide on elastomeric hemisphere is measured using a low power
microscope.

2.2 Determination of Young's Modulus, E

Determination of E, although always following measurements of a in
experimental order, will be presented now since this knowledge is essential
for the exploitation of equation 3. Having completed all the experiments
deemed necessary concerning contact area of the small flat in contact with
the elastomer, a large heavy disc of glass is placed on the hemisphere and
the contact radius (much larger than that for the small flat) determined.
This was done for glass discs of ca. 50gms and ca. 100gms. By applying
equation 3 and eliminating W, E can be calculated directly. Two comments
should be added. Firstly, the reason for adopting this procedure after
experiments involving the small flat are completed is to reduce the risk of
introducing errors by treating a potentially irreversibly deformed
elastomer (due to the considerable pressures incurred by the heavy discs).
Secondly, Young's modulus is determined by this method, rather than, say,

by considering stress/strain measurements on a similar material, in order
for the actual value of E of the top of the hemisphere itself to be established. Using a second specimen of the material could easily introduce
errors of reproducibility.

2.3 Materials and Evaluation of W as a Function of M_c

For reasons of convenience and availability, rather than choice, the
present study was effected using a common elastomer, a styrene butadiene
rubber (SBR). The copolymer in question consisted of 40% styrene and 60%
butadiene. Dicumyl peroxide (DCP) was employed as the crosslinking agent.
By incorporating different amounts of DCP in the elastomer before milling,
it was possible to obtain, by moulding at 150°C, hemispheres of differing
degrees of crosslinking. In principle, degree of crosslinking increases
directly with the quantity of DCP incorporated. But in practice, correlation can be poor. As a consequence, once all mechanical experiments
(evaluation of a and E) had been finished for a given hemisphere, it was
'decapitated'. The upper part corresponding to the contact zone was
divided into three parts and swollen in toluene. By using the equilibrium
swelling figures and applying the Flory-Rehner relationship [11], the
average inter-crosslink molecular weight, M_c, was determined.

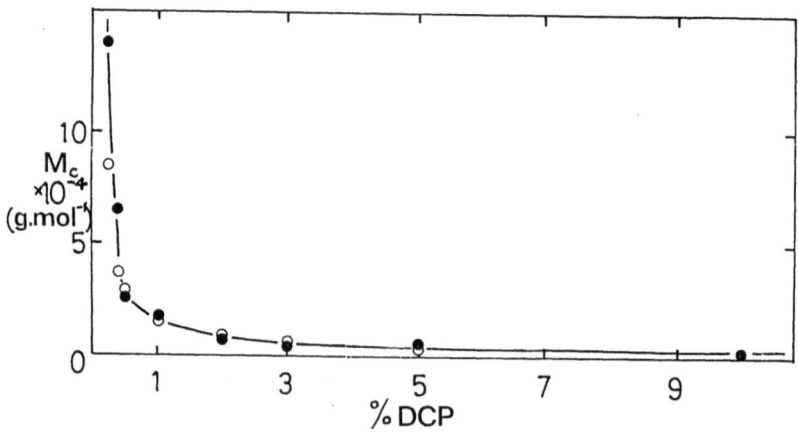

Figure 2. Average inter-crosslink molecular weight, Mc, as a function of
percentage DCP incorporated in SBR. Black and white points
represent respectively hemispheres used for SBR/glass and for
SBR/PMMA adhesion.

Fig. 2 gives the determined values of M_C as a function of percentage DCP (by weight) incorporated in the SBR.

Contact experiments were effected using SBR hemispheres of radius, R = 18.5mm, and small flat slides of glass and of polymethyl methacrylate (PMMA). Two types of contact were considered. In the first, referred to as "forced adhesion", the flat was placed on the hemisphere and a weight of 50gm symmetrically placed on top. After 5 minutes, a contact area exceeding the equilibrium value was assumed. The weight was removed and the diminishing contact radius followed as a function of time until (apparent) equilibrium was attained. Equation 3 was employed to calculate W. In the second type of experiment, referred to as "touching-on adhesion", asculating contact is established between hemisphere and flat. After contact, the radius a increases towards (apparent) equilibrium. Again, equation 3 serves to estimate W. In both types of experiment, the kinetics of evolution towards (apparent) equilibrium has been followed, but in the present context, we are essentially interested in static values of W. Suffice it to say that in a general manner, apparent equilibrium is reached faster for highly crosslinked hemispheres than for those of greater M_C. This being true both for forced adhesion and for touching-on adhesion. This is perhaps not surprising considering that the viscoelastic nature of the elastomer tends to decrease with increasing crosslink density.

3.   RESULTS AND DISCUSSION

3.1   Young's Modulus

As explained in section 2.2, Young's modulus of the SBR crosslinked to various degrees was determined by the indirect method of measuring contact radii of heavy discs and using equation 3. In order to investigate the validity of this procedure and, at the same time, that of the determination of M_C by swelling in tolmuene, we consider in Fig. 3 the evolution of E as a function of M_C, both scales being logarithmic.

Despite experimental scatter a fairly linear relationship may be seen. The Flory-Rehner relationship for M_C is of doubtful accuracy for both very high and very low degrees of crosslinking. Eliminating as a consequence the two points corresponding to the extreme values of M_C and applying regression analysis, a slope of -0.7 ± 0.1 is obtained. This is in fairly good agreement with the theory of rubber elasticity. The elastic modulus of a rubber may be written [12]:

$$E = mkT/Nv \qquad \qquad . \quad . \quad . \quad 4$$

where N equals the average number of monomer units, each of volume v, between crosslinks, kT represents Boltzmann's constant multiplied by absolute temperature and m is a numerical constant of order 1. Clearly N and M_c are directly proportional and we should thus expect a relationship of the form:

$$\log E \sim -\log M_c \qquad \qquad . \quad . \quad . \quad 5$$

to apply to Fig. 3, ie a gradient of -1. The linearity is present if the slope is somewhat feeble. This latter remark could be explained by the fact, previously mentioned, that the Flory-Rehner relationship is dubious at extreme values of M_c. It is quite possible that values of M_c approaching 10^5g.mol^{-1} have been slightly overestimated. Nevertheless, the correlation is considered as satisfactory evidence for the validity of the method employed for the determination of both E and M_c.

3.2 Adhesion and Crosslinking

    Having established values of E, equation 3 was employed to evaluate the apparent work of adhesion corresponding to both forced adhesion (W_I) and touching-on adhesion (W_c) at (observed) equilibrium. The values obtained for the SBR/glass system are given in Table 1 and those corresponding to SBR/PMMA in Table 2. As previously stated, values corresponding to very high and to very low M_c must be taken with reserve, given the limitations of the Flory-Rehner theory.

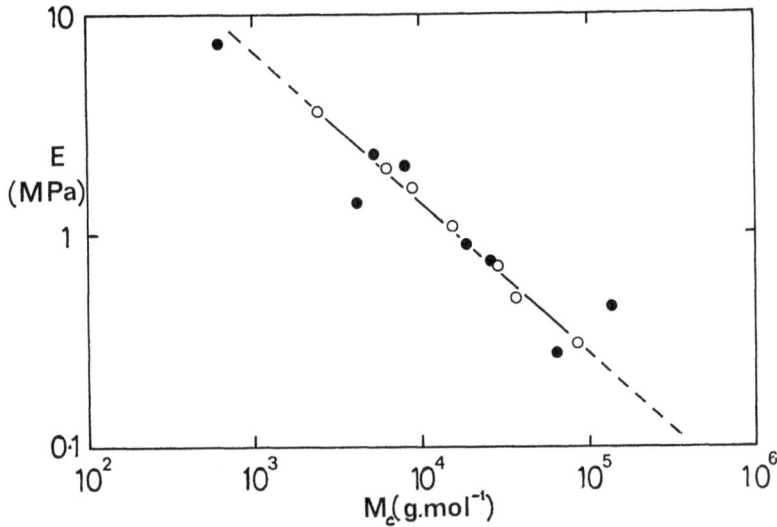

<u>Figure 3.</u> Logarithm of elastic modulus, E, vs log M_c. Symbols as in Fig. 2

TABLE 1
Values, of W_I and W_c as a function of M_c for the system SBR/glass

M_c(g.mol^{-1} x 10^{-3})	0.6	4	5.5	8.5	18.5	26.5	65	140
W_I(mJ.m^{-2})		54	51	114	99	77	340	210	580
W_c(mJ.m^{-2})		40	28	56	53	24	10	20	44

TABLE 2
Values of W_I and W_c as a function of M_c for the system SBR/PMMA

$M_c(\text{g.mol}^{-1} \times 10^{-3})$	2.5	6.5	9	16	29	37	85
$W_I(\text{mJ.m}^{-2})$		80	110	76	145	160	210	354
$W_c(\text{mJ.m}^{-2})$		73	50	46	64	56	42	55

Nevertheless, two observations can be clearly made for each of the systems studied:

  a) values of W_I increase markedly with M_c,
  b) W_c is generally fairly low and would seem to vary in a random manner
We shall therefore suppose that W_c is independent of M_c within limits of experimental error.

Forced adhesion experiments involve initial elastomeric strains in excess of those found at equilibrium. It is therefore conceivable that the pre-equilibrium states provoke some creep of the polymer leading to exaggerated values of contact radius, a, and therefore adhesion, W_I. Nevertheless, the reproducibility of the tests suggested that this is not the case and we therefore conclude that the increase of W_I with M_c is not an artefact, but a real correlation.

Despite the experimental error concerning W_c, the assumed independence of M_c allows us to calculate a mean value of touching-on adhesion, $\overline{W_c}$, for each system. In the case of SBR/glass we obtain $\overline{W_c} = 34 \pm 16\text{mJm-2}$ and for SBR/PMMA, $\overline{W_c} = 55 \pm 11\text{mJm}^{-2}$.

Using Fowke's expression for dispersive interactions [13], the reversible work of adhesion (given in equation 1) may be written as:

$$W_o = 2\left(\gamma_1^D \gamma_2^D\right)^{1/2} + I_{12}^P \qquad \cdot \quad \cdot \quad \cdot \quad 6$$

where γ_i^D represents the dispersive component of free surface energy $_i$, of material 1 or 2 in contact (SBR; glass or PMMA) and I_{12}^P is the polar interaction between the phases. Although debatable, we shall assume that I_{12}^p may be reasonably approximated by $2(\overset{p}{1}\overset{p}{2})^{1/2}$ where $\overset{p}{i}$ is the polar component of γ_i.

Taking $\gamma_{D}^{SBR} \sim 30 mJm^{-2}$, $\gamma_{P}^{SBR} \sim 1 mJm^{-2}$ [14], and $\gamma_{D}^{PMMA} \sim 36 mJm^{-2}$, $\gamma_{P}^{PMMA} \sim 4 mJm^{-2}$ [15], we obtain $W_o \sim 70 mJm^{-2}$. (The surface properties of SBR are little affected by its degree of crosslinking).

Since experiments were conducted at fairly high relative humidity ($\sim$50% RH) we may consider the glass slides (corresponding to a high energy material) to be covered with a layer of adsorbed water, for which $\gamma^D = 21.6 mJm^{-2}$ and $\gamma^P = 51 mJm^{-2}$. In this case we obtain $W_o \sim 65 mJm^{-2}$.

We may thus conclude that for both systems, $\overline{W}_c \sim W_o$. The order of magnitude is correct and although $\overline{W}_c < W_o$, this may well be attributable to imperfect molecular contact between the SBR hemispheres and the flat slides.

Carré and Roberts have previously suggested the equivalence of W_c and W_o[16]. The constancy of W_c suggest that the apparent work of adhesion evaluated in a touching-on experiment depends essentially on the surface properties of the materials in contact. Any molecular disturbances near the interface and especially near the "crack front" or ring defining the contact area, are presumably of minor importance.

In contrast to what occurs with W_c, forced adhesion is quite definitely related to M_c. Since W_c would seem to be approximately constant, we shall consider the increase in apparent adhesion with M_c above this value, ie. $W_I - W_c$.

Fig. 4 gives $W_I - W_c$ as a function of M_c for both systems; ordinate and abscissa being logarithmic scales. For the present elastomer, the number of carbon-carbon bonds between crosslinks, N, is ca $M_c/20$. N is also given in Fig. 4.

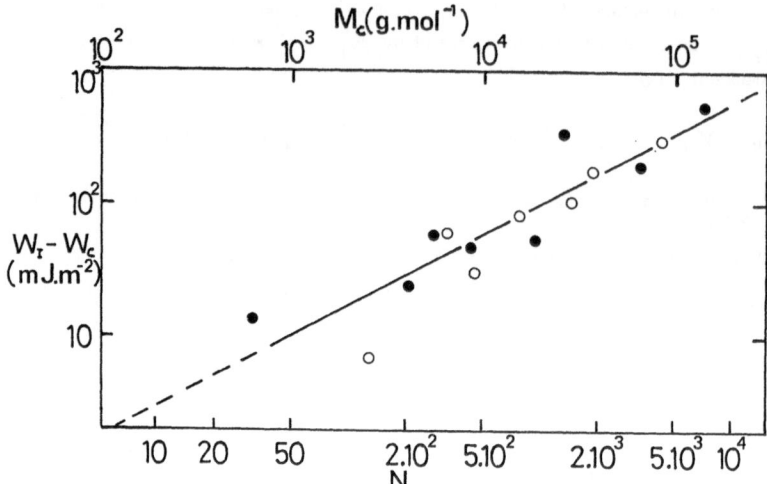

<u>Figure 4</u>. Evolution of log $(W_I - W_c)$ with log N (or log M_c) symbols as in Figs. 2 and 3

As can clearly be seen, experimental scatter is quite important. This at least partially linked to the elastomer studied since its structured form is rather poorly defined - particularly as far as the constancy of M_c is concerned for a given specimen. Nevertheless, the experimental points of Fig. 4 do seem to belong to the same family and regression analysis leads to a gradient of 0.8 $\pm$ 0.1.

A complete interpretation of this phenomenon is not yet available, but it is believed that the increase in apparent adhesion is closely related to either molecular dissipation [6,7] or to a chain extraction effect [8]. Since both processes are likely to show the same dependence on M_c or N, it would be rather difficult to choose between them given the imprecision of the results presented.

Let us first consider the possibility of molecular dissipation. During the failure, or separation process, molecular chains near the "crack front"

are atretched in the SBR. The length of these chains will be essentially
limited to N since, once we come across a crosslink in the mass, a more
rigid network is being considered. In effect we assume that the bulk
polymer in a crosslinked state is virtually unaffected by local,
essentially surface phenomena. It is thus only "free" chains bridging
the network and the rigid substrate which are stretched. This stretching
will lead to a certain quantity of energy being stored, proportional to N.
At failure, the stretching will disappear but the strain energy associated
will be rapidly dissipated. The actual interfacial separation involves,
in the present study, only physical bonds, but in fact all the chemical
bonds of the molecular chain of length N up to the first crosslink are
strained, even if only slightly. It is the associated energy that is
dissipated on failure. We can thus explain the decrease in apparent work
of adhesion, W, with increase in degree of crosslinking which is accompanied by a decrease in N.

Given the predicted proportionality between N and W_I-W_c, we should expect
the gradient of Fig. 4 to be 1. However, the simple model presupposes
that only points corresponding to chemical crosslinks define the average
value of N. No allowance is made for physical crosslinks being caused by
chain entanglements. For high degrees of crosslinking, physical
entanglement is likely to play a minor role but for the high values of M_c,
physical effects will presumably be important and reduce, at least to some
extent, the effective value of N. As a result, for high M_c, N effective
being swollen, the value of W_I-W_c could well be lower than predicted.
This could possibly explain the observed value of the gradient of Fig. 4
being less than 1.

The fact that W_I-W_c tends to a low value ($<5mJm^{-2}$) for low N (<20)
suggests that the effect of molecular dissipation disappears when crosslink
points become numerous. Elastomeric behaviour tends to become strictly
elastic without dissipation being important. This is in agreement with
the classic evolution of the complex modulus, E*, of an elastomer with
crosslinking degree.

Similar results have been observed with preliminary tests using a "beam
peel" method presently being developed.

In the second model, proposed by de Gennes [8], the measured energy of adhesion corresponds to the sum of the work required for chain scission and that required for chain extraction from the bulk environment. For the present study, the first effect, which corresponds more nearly to physical interfacial failure, will be masked by the second, equivalent to a van der Waals type bond energy multiplied by the number of monomer units in the chain segment between crosslink points, ie. N. Again, $W_I - W_c$ should be a linear function of N, or M_c, but for the same reasons as cited above, the gradient of Fig. 4 could be reduced by the presence of physical crosslinks caused by chain entanglements.

It is difficult to choose between these two proposed mechanisms to explain with any certainty the actual failure process, given the imprecision of the experimental results. Nevertheless, the fact that $W_I - W_c$ is highly dependent on N, or M_c, is an undeniable experimental observation.

4. CONCLUSIONS

a) Using the test method in which a small, flat, rigid slide is put into contact with an elastomeric hemisphere (method JKR), it has been found that the work of adhesion is virtually independent of the degree of elastomer crosslinking under conditions of touching-on adhesion.

b) In contrast, forced adhesion is closely related to crosslinking degree, in such a way that adhesion decreases with increasing crosslinking.

c) The difference between the values obtained for the two types of experiment would seem to be directly related either to a phenomena of energy dissipation at a molecular level or to a process of chain extraction close to the elastomer/rigid solid interface.

ACKNOWLEDGEMENTS

The authors wish to express their thanks to the Direction des Recherches, Etudes et Techniques (DRET), Ministry of Defence (France) for financial support and particularly for arranging the detachment to our laboratories of Philippe Schreck, whilst fulfilling his National Service.

REFERENCES

1. Dupré, A., 'Théorie mécanique de la chaleur', Paris, 1869.

2. Gent, A.N., Schultz, J., J. Adhesion, 3, 281, 1972.

3. Andrews, E.H., Kinloch, A.J., Proc. Roy. Soc. (Lond), A. 332, 385, 1973.

4. Maugis, D., Barquins, M., J. Phys. Appl. Phys., 11, 1989, 1978.

5. Williams, H.L., Landel, R.F., Ferry, J.D., J. Amer. Chem. Soc., 77, 3701, 1955.

6. Lake, G.J., Thomas, A.G., Proc. Roy. Soc. (Lond), A. 300, 108, 1967.

7. Carré, A., Schultz, J., J. Adhesion, 17, 135, 1984.

8. de Gennes, P.G., Adhésion: une liste de questions, 5th Mediterranean Summer School on Surface States and the Chemical Bond, 1983.

9. Johnson, K.L., Kendall, K., Roberts, A.D., Proc. Roy. Soc. (Lond), A. 324, 301, 1971.

10. Hertz, H., J. Rein Angew. Math., 92, 156, 1881.

11. Flory, P.J., 'Principles of Polymer Chemistry', Cornell University Press, ch. 13, 1971.

12. Treloar, L.R.G., 'The Physics of Rubber Elasticity', Clarendon Press, p. 66, 1949.

13. Fowkes, F.M., 'Treatise on Adhesion and Adhesives, Vol. 1', ed. Patrick, R.L., Marcel Decker, p. 344, 1967.

14. Carré, A., Thesis, Université de Haute Alsace, 1980.

15. Rance, D.G., 'Industrial Adhesion Problems', ed. Brewis, D.M., Briggs, D., Orbital Press, ch. 3, 1985.

16. Carré, A., Roberts, A.D., J. Chim. Phys., 83, 549, 1986.

5

ENERGY LOSSES IN PEELING OF UNVULCANIZED RUBBERS

K N G FULLER and G J LAKE
Malaysian Rubber Producers' Research Association
Brickendonbury, Hertford, Hertfordshire SG13 8NL, England

INTRODUCTION

The ability of unvulcanized rubber to adhere to itself, to other rubbers or to more rigid materials is exploited in such applications as rubbery adhesives and pressure sensitive tapes. It is also relied upon during the fabrication of rubber articles, including tyres, to keep the component parts in place prior to the article being vulcanized. This paper is concerned with the measurement of the adhesion between unvulcanized rubbers and their adhesion to more rigid substrates, rather than measurement of the bonds provided by adhesives or pressure sensitive tapes. The peel test is often used to determine the strength of the adhesion, and the degree to which peel measurements for unvulcanized rubber are affected by energy losses in the peel bend forms the main subject of the present work. The conditions under which a contact is produced affect the adhesion [1]; factors such as contact duration, pressure and temperature need to be controlled.

The work required, P, to separate unit area of the surface provides a definition of the strength of the adhesion, or the strength of an adhesive bond. Where one or both of the substrates are flexible the peel test (Fig.1) is a convenient means of measuring P. Adopting an energy balance approach, the peel energy can be easily related [2,3] to the force, F, applied to the peeled leg. For the T-peel geometry (Fig.1a)

$$P = 2F/w \tag{1}$$

where w is the width of the strip, and for the geometry in figure 1b,

$$P = F(1 - \cos\theta)/w \tag{2}$$

It has been assumed that all the energy is expended in the detachment process itself.

Two other mechanisms which can use up part of the work done are elastic extension of the peeled leg and irrecoverable deformation associated with the peel bend. For tapes and similar flexible but relatively inextensible substrates the first factor is often negligible.

(a)   **(b)**

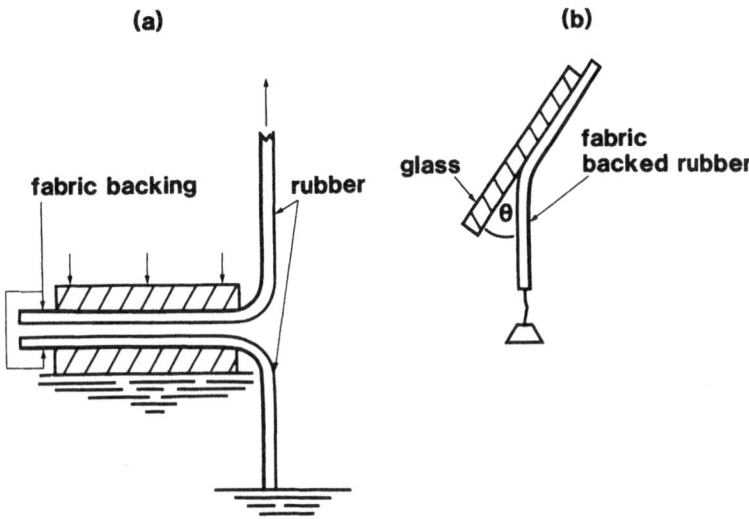

Figure 1. Schematic diagram of peel tests: (a) T-peel geometry for rubber-rubber adhesion, with pressure jig for producing contact prior to peeling. (b) geometry for rubber-'rigid' substrate tests at various peel angles, θ .

In the case of the peeling of rubber strips the same situation can be achieved by fixing a suitable backing to the strip. Even if the peeled leg does stretch significantly, a correction can be easily calculated provided the extension is predominantly elastic (4).

Two processes - plastic yielding and hysteresis - can result in energy losses in the peel bend. Several workers have analysed how yielding in the peel bend of an adherend influences the peel process by looking at its effect on the peel moment or the stresses induced in the adhesive layer (5,6,7). A later approach considered the energy associated with yielding; its magnitude was measured and found to compare well with calculated estimates (8,9). Hysteretic dissipation in the peel bend was discussed by Gent & Petrich (10) in the context of their experiments with a model viscoelastic adhesive. At high peel rates they found the peel force to differ between two types of peeling arrangement. It was much higher when the adhesive layer was peeled from the substrate and forced to bend through 90°, compared with the arrangement in which the substrate was peeled and the adhesive layer held flat. They attributed the difference to hysteretic losses in bending the adhesive layer through 90°; at high peel rates the adhesive was approaching its T_g so that the true strength of the adhesion - due mainly to flow within the adhesive - was low compared with the hysteretic losses.

Several investigators (1, 11) have used the peel test to measure the adhesion of unvulcanized rubber. Such materials are likely to show a high degree of hysteresis, and thus there is the possibility of significant viscoelastic losses as the strip passes through the peel bend. There appear, however, to have been few studies which have attempted to quantify such losses and see whether they contribute significantly to the observed peel force. They were invoked by Kendall (12) to explain the high apparent adhesion between an uroalkyd paint film and glass. He derived an expression relating the observed peel force to the ratio between the short- and long-time moduli of the film. The measurement of the cohesive strength of polymers by the tear test involves bending of the testpiece, and so the force to propagate the tear may be affected by hysteresis losses in the bend. This has been investigated for a series of semi-crystalline polymers (13) and for rubbery composite propellants (14); the contribution of the bending losses to the apparent tearing energy was observed to be in qualitative agreement with the predictions of a simple model, and in some circumstances to be highly significant.

Measurements showing the effect of the energy losses in the peel bend are reported here for unvulcanized natural rubber and epoxidized natural rubber peeled from a variety of substrates. The results are compared with estimates from simple calculations.

Two methods are used to detect the presence of energy losses not associated with the detachment process itelf. One, following Gent and Petrich (10), is to compare tests in which the substrate is held flat and the rubber peeled with those where the reverse procedure is adopted. The other is to investigate the variation of the rate of peel with the angle of peel for a constant apparent peel energy, P, as calculated from the peel force using equation (2). The equivalent experiment where the variation of the peel force at a given rate of peel is measure as a function of peel angle has been reported for pressure sensitive tapes (15, 16, 17) and for various polymer films peeled from glass (3,18). The tapes showed departure from the expected $F \propto (1-\cos\theta)^{-1}$ behaviour at high angles, possibly due to inelastic buckling in the backing (15). At low angles there was evidence of a transition from cleavage to shear failure in the adhesive (16). Apart from the experiments reported by Hata (18), there appears to have been little investigation of the effect of angle on the force for the peeling of unvulcanized elastomers.

EXPERIMENTAL DETAILS

Materials

The elastomers used were natural rubber (NR;SMR L) and 50% epoxidized natural rubber (ENR); in the latter 50% of the original double bonds are replaced by epoxide groups, thus raising the T_g from $-70°C$ to about $-20°C$. The rubbers were masticated to a set Mooney viscosity $[M_L(1+4)$, a measure of the shear viscosity at $100°C$ (see ISO standard 289)] in the range of 25 to 70 units, and moulded into sheets at $100°C$ or $130°C$ between aluminium foil or polyethylene terephthalate (PET, Mylar) film. Two procedures - degassing the rubber after mastication and cooling the mould under pressure - were tried in a partially successful attempt to reduce the formation of bubbles in the moulded sheets, a problem particularly with the epoxidized natural rubber and the low viscosity natural rubbers. Test strips 1 or 2.5cm wide and about 15cm long were cut from the sheets.

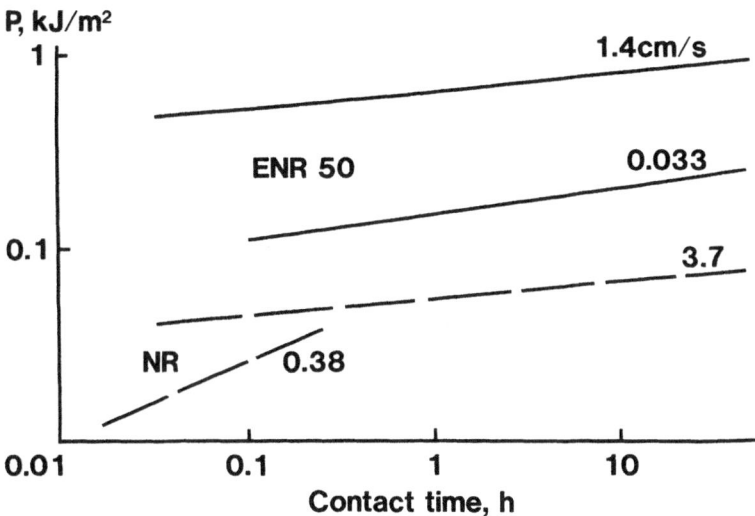

Figure 2. Apparent peel energy, P, as function of contact time at different peel rates for NR (M_L-26) or ENR (M_L-71) adhered to glass. Contact made by finger pressure.

When peeling a rubber strip from the substrate the foil or PET film on one side of the strip was removed and replaced by masking tape. The tape prevented deformation of the peeled part sufficiently for it to be regarded as inextensible. For some experiments in which the peeled pair was rubber and PET film the measurements were carried out on the as-moulded interface. Otherwise, the foil or PET film used in moulding was removed from the other surface of the rubber strip, and that surface brought into contact with the required substrate under a controlled pressure. Generally a piston device driven by compressed gas was used to hold the surfaces together for a known time; for some tests the surfaces were lightly brought into contact with finger pressure and left in contact for a known period. The latter technique gave results as reproducible as the piston device.

The substrates used were glass, polyethylene film or PET film; the rubbers were also peeled from each other.

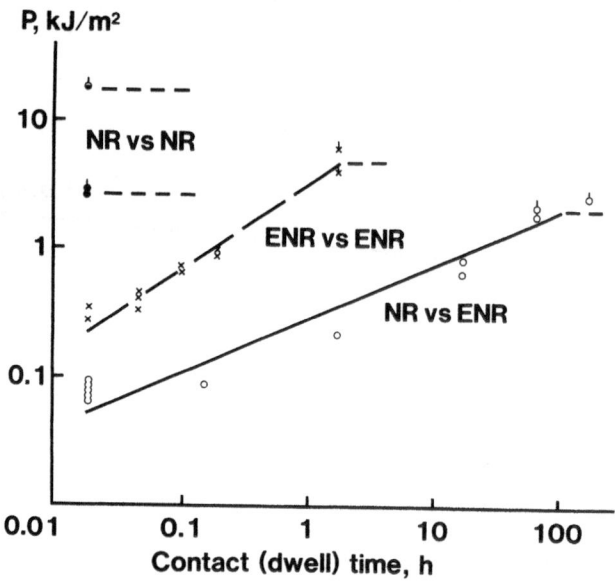

Figure 3. Apparent peel energy, P, as function of contact time for various rubber systems. Contact pressure: 0.15MPa. Adhesive failure is indicated by a flag thus .

Test Procedure

Most of the measurements were carried out using the peel geometry shown in figure 1b; here a flexible strip is peeled from a rigid substrate. Polymer film or rubber could be held flat and thus made to act as the rigid substrate by fixing it to the glass plate with double-sided adhesive tape. A weight was applied to the strip to be peeled, and the rate of peel measured. Rates too rapid to be determined by timing directly with a stopwatch were obtained from video recordings.

Some tests involving the peeling of the two rubbers were performed using the T-peel geometry shown schematically in figure 1a. After bringing the strips into contact for the desired time, the pressure was released and the strips peeled in an Instron testing-machine, at a crosshead speed of 200mm/min (equivalent to a peel rate of 100mm/min) unless otherwise indicated. The peel force was output to a chart recorder and an average estimated.

The peel measurements were all made at ambient temperature (~21°C). The surfaces were generally brought into contact at the same temperature, but in a few instances the as-moulded interface was used.

67

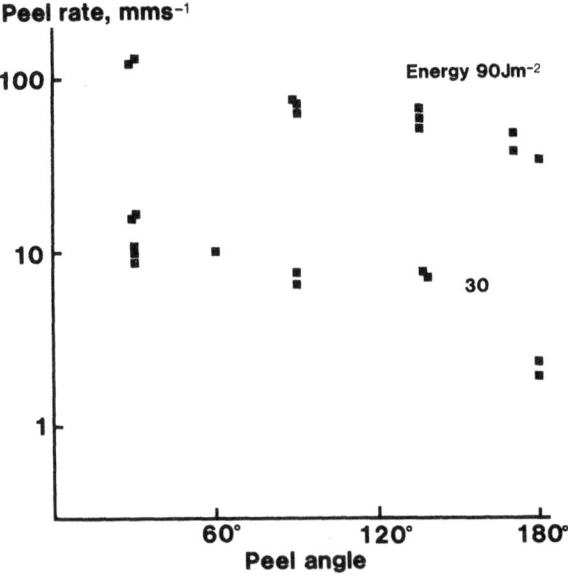

Figure 4. Rate of peeling of NR (M⨯ 30, thickness 2mm) from glass as
function of peel angle for two constant apparent peel energies. Contact
by finger pressure; contact time 2min.

RESULTS

The contact conditions – in particular time and pressure – were chosen in
the light of a previous study of the mutual and self-adhesion of ENR and
NR and their adhesion to glass (19). Results from that study are given in
figures 2 and 3. It is seen that for the rubber-glass system (Fig.2) the
apparent peel energy increases quite slowly with contact time
particularly, after a few minutes, on a linear time basis. The different
behaviour of NR at low peel rates was associated with a tendency to flow –
a situation avoided in most of the present tests. It is interesting to
note that the increase of the adhesion, though slow, continues for a long
period. For the mutual adhesion of NR and ENR (Fig.3), the change of the
apparent peel energy with time is rather greater. Moreover, because the
relation between the apparent peel energy and rate is very steep
(approximately, rate α (energy)^8 (19)), the rate at a given energy
varies rapidly. Although the mutual adhesion of the rubbers increases more
rapidly than the rubber-glass adhesion, the time-scale necessary for the
adhesion to reach its full extent is much longer than that required for the
self-adhesion of either rubber.

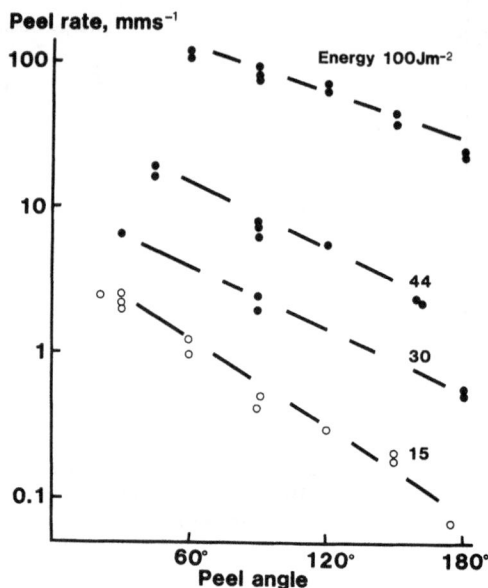

Figure 5. Rate of peeling of ENR (M_L - 24, thickness, 2mm) from glass as a function of peel angle for a range of constant apparent peel energies. Contact by finger pressure, contact time 2min.

The contact time chosen for the present tests was 2-3 minutes. This was conveniently short, yet sufficient for the additional contact time during peeling to have been insignificant even for the slowest peel rates.

The effect of contact pressure was such that above 0.1MPa the increase of the adhesion with pressure was modest. Pressures of 0.1 to 0.2MPa were adopted as standard for the pressure jig. The relative insensitivity except at very low contact pressures probably accounts for the consistency of the results from tests using finger pressure to adhere a strip.

The rate of peeling of NR from a glass substrate is shown in Figure 4 as a function of peel angle for two constant apparent peel energies. The rate in each case is seen to decrease about fourfold as the angle rises from 30° to 180°. If all the work of peeling had gone only into the detachment process itself, the rate should have remained constant unless some change in the mode of detachment had occurred as the angle altered. The results of similar experiments for the peeling of ENR from glass (Fig.5) show a somewhat larger fall in the rate with increasing angle;

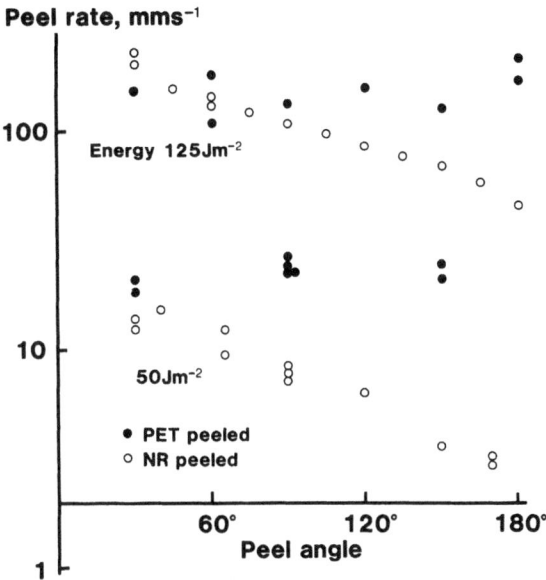

Figure 6. Peeling of NR (M_n-30) and PET film as a function of peel angle with one member of the pair held flat and the other peeled. For data at apparent peel energy of $125Jm^{-2}$: contact pressure 0.2MPa for 3min., rubber thickness 0.5mm; for energy $50Jm^{-2}$: contact by finger pressure, contact time 2min., rubber thickness 2mm.

the extent of the fall is broadly similar over the range of energies studied.

The association of the change of rate seen in figures 4 and 5 with the bending of the rubber strip is supported by the data of figure 6. Here the adhesive pair was natural rubber and PET film. When the former was the peeled member   the rate of peeling again decreased about fourfold as the peel angle was raised. If, however, the natural rubber was held flat and the PET strip peeled, the rate did not change with increase of the peel angle. The rate observed for peeling the PET is seen to be comparable with the value observed at low angles ($<60°$) for peeling the natural rubber. It thus appears that the latter procedure gave incorrect data at high peel angles, which was probably associated with the bending of the rubber strip. Although the PET film appeared to behave in an ideally elastic manner over the range of peel energies (up to $125Jm^{-2}$) investigated here, other workers (9) have found that at higher peel energies ($-1000Jm^{-2}$) this material can yield, and thus introduce losses in the peel bend. The constant rates observed for the

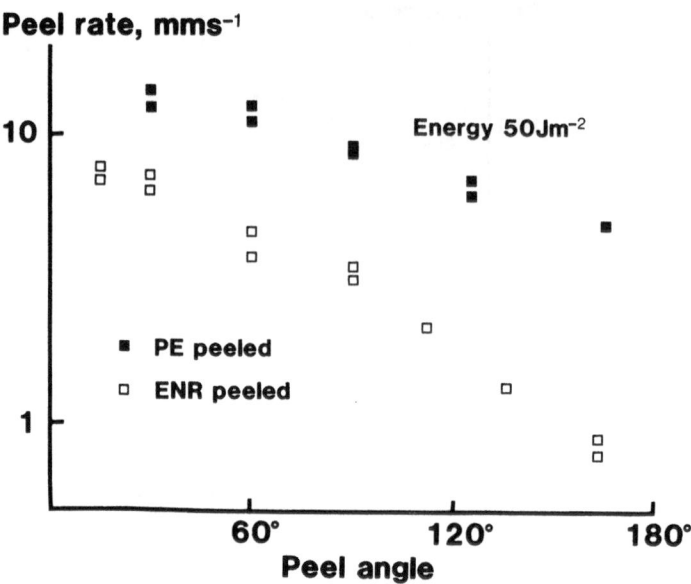

Figure 7. Peeling of ENR (M_L-24) and polyethylene (PE) film as a function of peel angle with one member of the pair held flat whilst the other was peeled. Contact by finger pressure; contact time 2min; rubber thickness 2mm.

peeling of the PET film indicate that at least for angles down to 30°, there was no obvious transition in the mode of failure of the interface as found by Kaelble for some systems (16).

Figure 7 shows peel rate data at one apparent peel energy for epoxidized natural rubber in contact with polyethylene film; PET film could not be used in combination with ENR because the adhesion was too strong for interfacial failure to be produced. Peeling of the ENR gave rates which decreased by about an order of magnitude as the angle was raised - a change similar to that in figure 5 for the ENR-glass system. The polyethylene when peeled from the ENR did not, unlike the PET, give a rate independent of angle, presumably because the polyethylene had a much lower yield stress - 8MPa compared with 110MPa. The two sets of data again tend to converge at low angles.

The results presented in Figures 4 to 7 show that in peeling an unvulcanized rubber from a substrate there can be a significant decrease in the peel rate at a constant apparent energy as the angle is raised. For tests in which the peel force is measured for a constant peel rate the

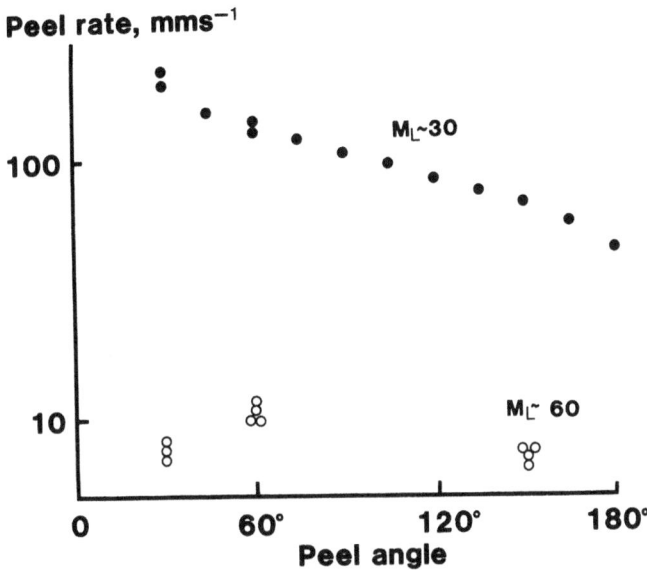

Figure 8. Effect of rubber viscosity upon the angular dependence of the rate of peeling of NR from PET film. Points ●: apparent peel energy 125Jm^{-2}; contact presure 0.2MPa for 3min. Points ○: apparent peel energy 40Jm^{-2}; interface as moulded. Rubber thicknesses, 0.5mm.

changes would be expected to be proportionately smaller, though no less significant, because the rate generally rises rapidly with the peel energy. The data for a range of energies for peeling ENR from glass (Fig.5) indicate that the roughly tenfold decrease in the rate as the angle rises from 30° to 180° would correspond to a threefold increase in the apparent peel energy.

The extent of the angular dependence of the peel rate has been seen to differ for ENR and NR; the viscosity of the rubbers differed only slightly, but the former has a much higher T_g. The influence of the viscosity on the dependence observed for one rubber – NR – is shown in figure 8, where results for peeling strips of M_L-60 are contrasted with those of M_L-30. The higher viscosity NR gave no angular dependence for energies ranging up to at least 50Jm^{-2}.

The possible influence of the thickness of the peeled rubber strip on the angular dependence of the rate was investigated by measuring the rate at a fixed, high peel angle (150°) and a constant apparent peel energy over a range of strip thicknesses. This was produced by laminating up

72

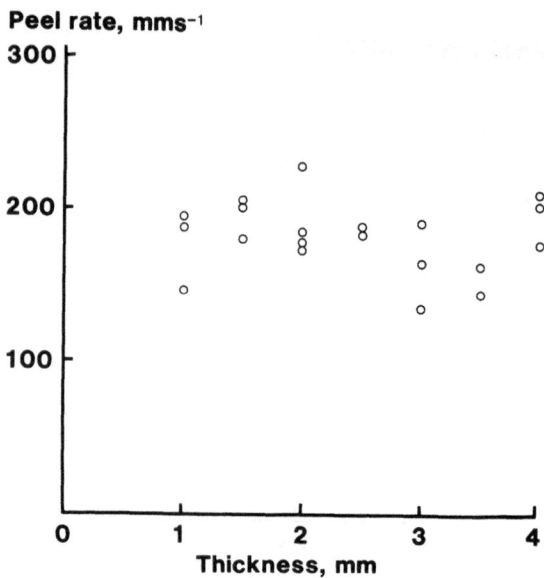

Figure 9. Effect of thickness of peeled rubber strip upon peel rate at a peel angle of 150° and an apparent peel energy of 250Jm^{-2}. NR (M_L-30) peeled from PET film. Contact pressure 0.2MPa for 3min.

0.5mm thick NR (M_L-30); the high autohesion ensured that no delamination occurred and the technique had the advantage that all the peel interfaces were formed from rubber of one sheet, thus improving reproducibility. The data (Fig.9) indicated no obvious effect of thickness. According to the data already presented such a rubber peeled at this high angle would give a rate much reduced by losses in the peel bend; the losses must therefore be independent of strip thickness.

The markedly different angular dependence seen for ENR and NR supports the contention that energy is absorbed in the bulk of a peeled strip. The strips were all backed with masking-tape to reduce the extensibility of the peeled part, and it is possible that energy may have been absorbed by the tape and its adhesive as it passed through the peel bend. The fact that the higher viscosity NR showed no angular dependence of the rate (Fig.8) suggests, at least in that case, that the tape did not contribute to energy losses within the peel bend. The data in that figure were obtained with thin (0.5mm thickness) strips. With thicker rubber strips another deformation mechanism involving the tape was observed. This was buckling of the tape due to the compressive forces on the inside of the peel bend, a phenomenon noted by another worker in similar circumstances (15). The neutral axis in the bend was apparently

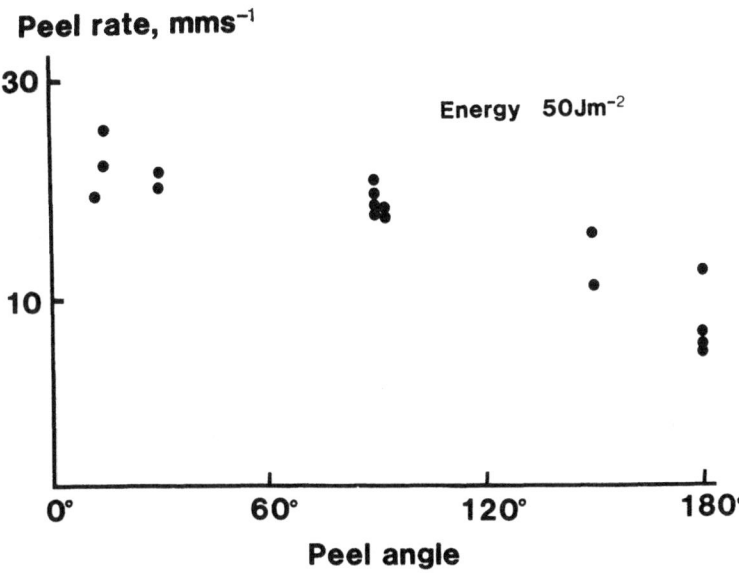

Figure 10. Angular dependence of rate for PET film backed with masking-tape and peeled from NR (M_L-30) at an apparent energy of 50Jm^{-2}. Rubber thickness 2mm; contact by finger pressure; contact time 2min.

located somewhere within the rubber and for larger peel angles and thicker strips the compressive force was sufficient to buckle the masking-tape. It appears, however, that the buckling did not significantly effect the peel bend losses as there was no difference in the rate of peel of rubber strips as the thickness varied (Fig.9). As an additional check on the influence of the masking-tape, it was attached as a backing to strips of PET film, which were peeled from NR over a range of peel angles. The results (Fig. 10) show a slight angular dependence of the rate. Since the PET film alone gave no angular dependence, the change seen in the figure must be ascribed to the masking-tape, in apparent contradiction with the earlier conclusions on the influence of the tape. It was observed, however, that bands of detachment were produced as the tape-backed film passed through the peel bend. The tape and rubber had appeared to maintain complete contact even in regions of buckling. Thus the energy absorbed in the process of detachment could account for the angular dependence of the peel rate for the tape-backed PET film.

The angular dependence of the peel rate for natural rubber in contact with epoxidized natural rubber is shown in figure 11. The test geometry used was that shown in figure 1b so that one of the rubber strips could be held flat by fixing it to the glass plate and the other rubber strip

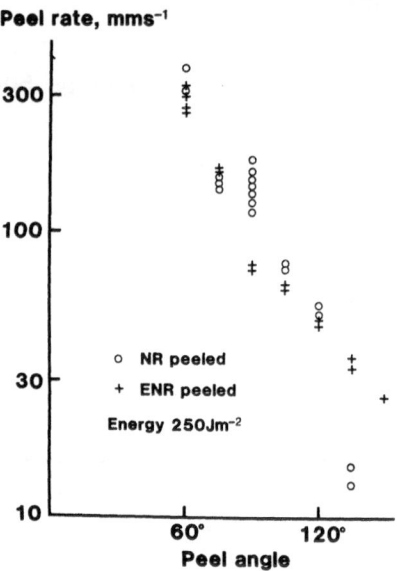

Figure 11. Angular dependence of peel rate for NR adhered to ENR with one strip held flat and the other peeled. Rubber viscosities, M_L-40; thicknesses, 2mm. Apparent peel energy, 250Jm^{-2}. Contact pressure, 0.2MPa for 3min.

peeled at a series of angles. The range of angles covered is limited because at low angles shear flow through the thickness of the strips could occur and at high angles bulk flow in the peeled leg, in the case of NR, could take place. Between 60° and 120° the rate is seen to fall by about an order of magnitude for both rubbers in this particular test. The fall with angle is greater than that observed for rubber peeled from the other substrates, perhaps because the peel force is larger. With both peeled materials giving a strong angular dependence, the magnitude of the rate in the absence of peel bend losses cannot be estimated with any accuracy. It does appear that the peel force obtained from the T-peel geometry for these rubbers can be significantly increased by the losses in the peel bends. The error in the force can be estimated from the relationship between the apparent peel energy (αforce) and the peel rate for the upper limit of the mutual adhesion of NR and ENR; the rate was found to vary as approximately (energy)[8] (19). This steep dependence means that an order or magnitude reduction in the peel rate due to peel bend losses would be equivalent to an increase in the peel force of about one-third.

DISCUSSION

The angular dependence of the peel rate observed for rubber strips
suggests that energy losses in the peel bend are small at low peel angles,
steadily increase with increasing angle, are independent of the thickness
of the strip and depend upon the hysteretic properties of the rubber.
These observations can be compared with the predictions of two simple
models.

If the rubber is assumed to be a linearly elastic material, bending
theory can be applied to the peeled strip (20, 21). The energy, U, stored
per unit length in the bend,

$$U = (Fy)^2/2EI$$

where F is the peel force, y the moment arm of the peel force about the
peel front, E is Young's modulus and I the moment of area of the strip
about its neutral axis. From bending theory

$$y^2 = \frac{2EI}{F} (1 - \cos\theta)$$

thus

$$U = F(1 - \cos\theta).$$

If a fraction, H, of this energy is lost in the peel bend, equation (2)
can be modified so that the apparent peel energy, P_a is given by

$$P_a = P/(1-H) \qquad\qquad (3)$$

where P is now the true peel energy. The relation between P and P_a is
the same as that given by Gent & Jeong (13) for the true and apparent
tearing energies.

Although equation (3) correctly predicts that P_a is independent of
the thickness and varies with the hysteretic factor H, it suggests that
P_a, and hence the peel rate at a given P_a, does not change with the
peel angle. In this important respect the model completely fails to fit
the observations. Measurements of the hysteretic factor, H from

TABLE 1

Hysteretic factor of rubbers

Rubber	M_L	H
NR	30	0.38
NR	60	0.25
ENR	40	0.38

extension and retraction tests at a strain rate of 2min^{-1} (see Table 1)

show that the angular dependence of the peel rate does not correlate
simply with 1-H when the results for the ENR (M_L - 40) and NR (M_L - 30)
are compared. The absolute magnitudes of the change also seem to be
rather larger, certainly by 180°, than is predicted from the values of H
in Table 1. The figures would be strongly influenced by the strain rate
but this is likely to be higher in the peel tests and thus H lower. Gent
and Jeong (13) similarly did not find a quantitative correlation with H;
all their data concerned tear tests where the strips were bent through
90°, so no data on the effect of angle were presented.

It seems likely that the strains near the peel front are sufficiently
high for the assumption of linear behaviour to fail, and the values of H
could be high enough to require the viscoelastic nature of the rubber to
be taken into account in analysing the geometry of the peeled strip.

An alternative estimate of the energy losses can be made if a
radius of curvature, R, at the peel front is measured directly or assumed;
such an approach is similar to that adopted previously to estimate losses
associated with plastic yielding (9,22). The elastic energy, U, stored
per unit length in a bend of radius R,

$$U = kEt^3w/6R^2 \qquad (4)$$

where t is the thickness and w the width of the strip, and k is a constant
dependent on the location of the neutral axis. If the backing were
incompressible along its length, the neutral axis would be located at the
backing and k equal to 1. In practice the neutral axis is likely to be
located somewhere between the inside surface and the centre of the rubber
strip, and thus k range from 1 to 0.25. Observation showed the form of
the peeled strip did not appear to follow the shape predicted by elastic
bending theory. The radius varied markedly near the peel front and thus
it was rather difficult to measure its value with any accuracy. A rough
estimate of the maximum energy losses possible in the bend can be obtained
by assuming the radius to be of the order of t and assigning a value of
0.5 to the constant k. For a hysteretic factor of 0.4, a thickness of
of 2mm and a modulus of 0.3 MPa (from the extension test of the NR (M_L-30)
at a strain rate of $2min^{-1}$) the energy loss per unit area of a strip is
50 Jm^{-2}. This figure is comparable with the change in the apparent peel
energy expected to produce the variation of peel rate with angle seen in
figure 6, but rather less than that expected from figure 11. Thus the
assumption that the angular dependence of the peel rate is predominantly
due to hysteretic losses in the peel bend appears to be borne out.

Attempts were made to compare estimates of the losses using equation
(4) with measurements of the work required to propagate a bend with a
controlled radius through a strip of unvulcanized rubber. Gent and Hamed
(9) devised an experiment to determine the energy to propagate a bend
through a backing which plastically deformed. Strips of the backing
material were constrained between vertical plates and the bend moved
through the strips by raising one of the plates. In the present case a
similar arrangement was found to give rather erratic results partly due to
the tendency of the straight portion of the rubber strip (which could not
be adhered to the rigid plates in this test) to buckle, rather than travel
through the bend, when the plate was moved.

As an angular dependence of the peel rate for unvulcanized rubber, associated with losses in the peel bend, occurs, the use of a backing with a high bending stiffness is suggested. The backing should not be susceptible to viscoelastic losses or plastic yielding, so a suitable choice appears to be spring steel. Preliminary measurements with such a backing, adhered to the rubber by means of double-sided tape, indicate that the angular dependence, over a range of peel angles from (30° to 120°), is absent, presumably because the maximum curvature in the peel bend is much decreased. There is also a significant reduction in magnitude of the peel force as expected from the data in figure 11.

CONCLUSIONS

Significant losses can occur in the peel bend when measuring the adhesion of unvulcanized rubbers by means of the peel test. The effect is accentuated at high peel angles. It appears that the losses can be much reduced by choosing a relatively inflexible backing, such as spring steel, which itself does not introduce energy loss processes like yielding.

ACKNOWLEDGEMENTS

We thank A.G. Thomas for helpful discussions and R. McKenzie, S. Micallef and D. Archer for assistance with the experiments.

REFERENCES

1. Hamed, G.R., Tack and green strength of elastomeric materials. Rubber Chem. Technol., 1981, 54, 576-595.

2. Rivlin, R.S., The effective work of adhesion. Paint Technology, 1944, 9, 1-4.

3. Hata, T., Mechanics of stripping. I. Relation between the angle and load of stripping. Chem. High Polymer (Japan), 1947, 4, 67-72.

4. Lindley, P.B., Ozone attack at a rubber-metal bond. Journal Inst. Rubber Ind., 1971, 5, 243-248.

5. Spies, G.J., The peeling test on Redux-bonded joints. Aircraft Engineering, 1953, 25, 64-70.

6. Mylonas, C., Analysis of peeling. In Proceedings of the Fourth International Congress on Rheology, Part 2, ed. E.H. Lee, Interscience, New York, 1965, pp 423-447.

7. Duke, A.J. and Stanbridge, R.P., Cleavage behaviour of bonds made with adherends capable of plastic yield. J. Appl. Polym Sci., 1968, 12, 1487-1503.

78

8. Gent, A.N. and Schultz, J., Equilibrium and non-equilibrium aspects of the strength of adhesion of viscoelastic materials. In Proceedings of Internatinal Rubber Conference, Brighton, 1972, Paper C1.

9. Gent, A.N. and Hamed, G.R., Peel mechanics for an elastic-plastic adherend. J. Appl. Polym. Sci., 1977, 21, 2817-2831.

10. Gent, A.N. and Petrich, R.P., Adhesion of viscoelastic materials to rigid substrates. Proc. Roy. Soc. A, 1969, 310, 433-448.

11. Bhaumik, T.K., Gupta, B.R. and Bhowmick, A.K., Tack and green strength of blends of bromobutyl and EPDM rubbers. I. Unfilled gum blends. J. Adhesion, 1987, 24, 183-198.

12. Kendall, K., Peel adhesion of solid films - The surface and bulk effects. J. Adhesion, 1973, 5, 179-202.

13. Gent, A.N. and Jeong, J., Contribution of bending energy losses to the apparent tear energy. International J. Fract., 1985, 29, 157-168.

14. Kinloch, A.J. and Tod, D.A., A new approach to crack growth in rubbery composite propellants. Propellants, Explosives, Pyrotechnics, 1984, 9, 48-55.

15. Kaelble, D.H., Theory and analysis of peel adhesion: Mechanisms and mechanics. Trans. Soc. Rheol., 1959, 3, 161-180.

16. Kaelble, D.H., Theory and analysis of peel adhesion: Bond stresses and distributions. Trans. Soc. Rheol., 1960, 4, 45-73.

17. Aubrey, D.W., Jackson, T.A. and Smith, J.D., Peel adhesion of pressure-sensitive tapes: Determination of peel adhesion with variation of the angle of peeling. Journal Inst. Rubber Ind., 1969, 3, 265-269.

18. Hata, T., Mechanics of stripping. II. Experimental examination of the theoretical formula of stripping. Chem. High Polymer (Japan), 1947, 4, 72-77.

19. Fuller, K.N.G., and Lake, G.J., Adhesion of unvulcanized rubber to various materials. In Adhesion '87, PRI, London, Paper X.

20. Kendall, K., The shapes of peeling solid films. J. Adhesion, 1973, 5, 105-117.

21. Gent, A.N. and Hamed, G.R., Peel mechanics. J. Adhesion, 1975, 7, 91-95.

22. Hatzinikolaou, T.A. and Packham, D.E., The interaction of bulk and surface properties in the peeling of some ethylene-vinyl acetate copolymers. In Adhesion '87, PRI, London, Paper 8.

6

ADHESION OF RUBBER TO RIGID SUBSTRATES: MEMORY EFFECTS AND THE THRESHOLD ENERGY

K N G FULLER and G J LAKE
Malaysian Rubber Producers' Research Association,
Brickendonbury, Hertford, Hertfordshire SG13 8NL, England

INTRODUCTION

Adhesion between rubber and more rigid materials is of importance for tyres and many other rubber articles. Although there have been notable advances in recent years, the mechanics of adhesion failure is still far from fully understood. Practical systems are often difficult to study in this respect because of complications such as uncertainty about the locus of failure. In the present studies a system in which vulcanized rubber is simply placed in contact with a rigid substrate (glass or coated glass) is being used as a model. This gives essentially interfacial failure. An energetics approach has proved very helpful in the study of various cohesive failure properties of rubber [1]. In earlier work with the present adhesion system it was shown, through peel tests at various angles, that an energetics approach could be used to describe the peel behaviour with the results being independent of peel angle provided this was 30° or more. At lower angles the peel energy was found to increase with decrease in angle. This was attributed to energy losses due to elastic instability in the peel front zone arising from extension of the peeled leg and sometimes giving rise to the occurrence of waves of detachment and reattachment [2,3].

In the present work, all experiments have been carried out at peel angles of 30° or more to avoid the above effects. Even with this precaution and the apparently very simple adhesion system adopted complications remain, one being that the strength of adhesion increases with the time of contact or 'dwell time'. Various explanations fo the dwell effect on rigid substrates have been proposed involving stress relaxation, impurity diffusion or chemical reaction, but the cause remains uncertain [3-7]. To try to clarify the mechanism, experiments are being performed with a highly purified, relatively inert rubber.

The existence of a threshold energy for the onset of cohesive failure in vulcanized rubbers is well-established and the observed value can be successfully predicted from the long-chain molecular structure and the primary bond strength [1]. A threshold energy (of different magnitude) would also be expected to apply to the detachment of rubber adhered to a rigid substrate, although its measurement can involve difficulties not present in the cohesive failure case. The experiments on the adhesion of vulcanized rubber to glass are being extended to measure the threshold energy. This involves long-term experiments and for such purpose a rubber purified by solvent extraction is highly desirable if not essential, as otherwise blooming of impurities is likely to occur at the interface and affect the adhesion. A relatively inert rubber is again virtually essential, as otherwise oxidation of the extracted material is likely to occur over the time-scales involved. Such oxidation can enhance the adhesion if it occurs whilst the rubber is in contact at the interface and, of course, such a system would be ill-defined and difficult to control for the present purposes. A theoretical estimate of the threshold energy of adhesion, taking into account the long-chain molecular structure of the rubber and the strength of the interfacial bonds, is compared with the experimental results.

EXPERIMENTAL PROCEDURE

The basic experimental method is illustrated schematically in Fig 1. A strip of rubber, normally about 1cm wide and 1mm thick, was carefully adhered to a glass plate mounted at a fixed angle Θ to the vertical and left in contact for a measured period. A constant force, F per unit width (referred to the unstrained state), was then applied to the lower end of the strip as illustrated and the rate of peel measured. The contact time was varied from a minimum of 2 minutes (adopted as a standard for short-term tests) up to a few years. The rubber used for most of the experiments was a copolymer of ethylene and propylene (Vistalon 404, Exxon Chemical Company) vulcanized by adding 2.7 parts per hundred rubber by weight phr of dicumyl peroxide and 0.32phr of sulphur and heating for 60 minutes at 160°C under pressure. The raw rubber was extracted before cure and the vulcanized sheet after cure using hot acetone, some 3% by weight of material being removed on each occasion. This procedure was necessary in order to prevent blooming of soluble impurities which can affect the adhesion, particularly over the long time-scales involved in many of the present experiments. Extracted rubber can be much more susceptible to oxidation and a relatively inert material was chosen on this account for the main experiments.

The rubber surface for which the adhesion was measured was moulded against glass. A separate glass plate was used for the adhesion measurements - this was soda-lime float glass, with an indium-tin oxide 'conducting' coating on one side. The test surface was cleaned with solvent (generally acetone) prior to each experiment. The peel rate was determined by timing the peel front movement with a stopwatch. Experiments were carried out on either the conducting or the non-conducting side of the glass at 22°C ($\pm$ about 2°C). Measurements were also made at 50°C, the plate being heated by passing a suitable current through the conducting layer and adhesion experiments being carried out only on the non-conducting side; the test piece was pre-heated to 50°C (in an oven) before it was adhered to the glass at this temperature to avoid thermally-induced stresses at the interface.

In order to avoid increases in peel energy that can occur at low angles, experiments were carried out at angles of 30° or more. In fact, because of the low energies involved with the present adhesion system, it

81

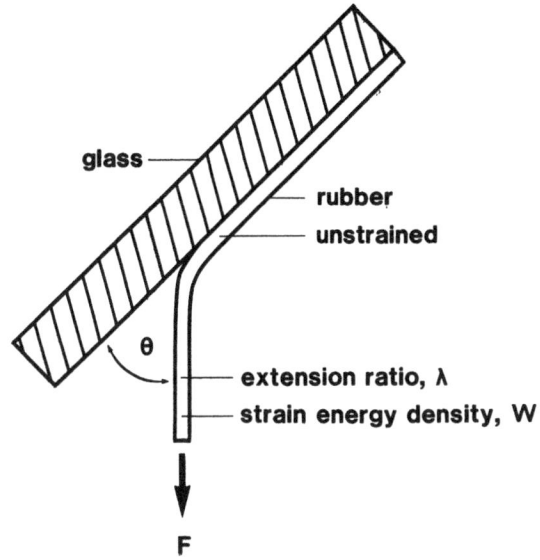

Figure 1. Peel test arrangement.

was convenient for many experiments to use an angle of 30°. However,
some experiments were also carried out at higher angles and no angular
dependence was observed over the range covered.

ENERGETICS ANALYSIS

For the case illustrated in Figure 1, where the adhered rubber is in the
undeformed state and the peeled rubber (away from the vicinity of the
peel front) is subject to a longitudinal extension ratio λ, it can
readily be shown that the energy available for peeling unit area of
surface is given by [8]

$$P = F(\lambda - \cos\theta) - Wt \qquad (1)$$

where F is the per unit unstrained width, W the strain energy density in
the peeled leg and t the unstrained thickness of the strip. If the
stress-strain behaviour of the material is linear, eqn. (1) becomes [9]

$$P = F\left(\frac{\lambda+1}{2} - \cos\theta\right). \qquad (2)$$

A simple extension stress-strain curve was carried out on the rubber
to enable the peel energy to be calculated from equation (1) or (2).

RESULTS

Typical results for the peel rate, at a given peel energy, against the
time for which the rubber and substrate were left in contact prior to
peeling are shown in Figure 2 for the extracted ethylene-propylene
rubber. An effect of this general nature is well known, both for
vulcanized rubber and for unvulcanized rubber adhering to various
materials [3-7]. The results in Figure 2 show quite a strong
dependence, perhaps reflecting the cleanliness of the rubber, the rate
decreasing by about 3 orders of magnitude over the time range covered
(some 4½ decades) and showing no signs of diminishing on the log-log
basis employed. In the vicinity of the energy used (8.5 J/m^2) the rate
is found to vary approximately as the cube of the energy for the
ethylene-propylene vulcanizate (cf Figure 6) so that the change in rate
shown in Figure 2 corresponds to an increase in the peel energy (for a
given rate) of about an order of magnitude.

In fact, from later observations the relationship indicated by the
results in Figure 2 appears to continue in similar fashion for at least
another decade of time (i.e. up to ca $2x10^4$h or approaching 2½ years).
Thus at the test temperature of ca 20°C the 'dwell' effect continues
over a very long period.

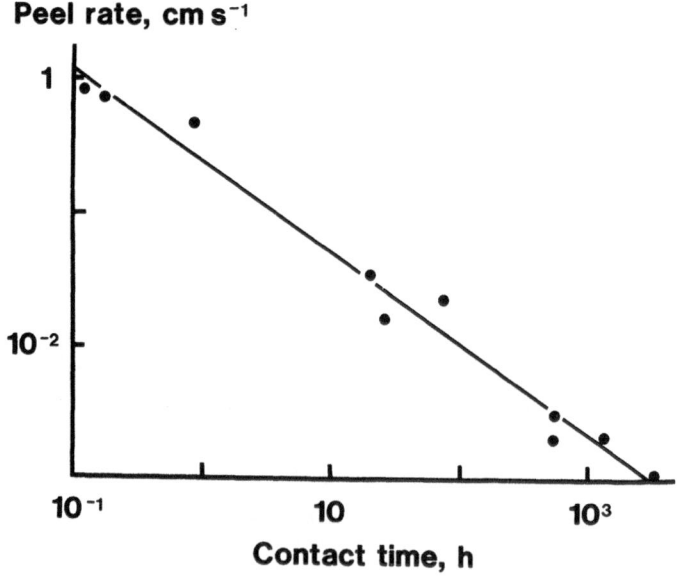

Figure 2. Peel rate versus contact time (logarithmic scales) for the
extracted vulcanizate of ethylene-propylene rubber versus conducting
(indium-tin oxide coated) glass for a constant peel energy (8.5J/m^2).

The dwell effect just discussed is well known but the effect
described below, which is associated with it, does not appear to have
been reported previously. Indications of the latter effect were first
obtained during some experiments on a vulcanizate of acrylonitrile –
butadiene rubber. A sequence of experiments was carried out in quick
succession using a standard contact time (0.1h) and the same (relatively
high) peel energy throughout. As the experiments proceeded, so the
adhesion achieved (during the 0.1h dwell) appeared to increase, the peel
rate tending to decrease slowly. The nitrile rubber was not extracted
and it appeared that the effect might be associated with a (tacky) bloom
although there was no direct evidence for this. As the decrease was slow
and the scatter in the measurements large, these experiments left the
significance of the effect in some doubt. However, further experiments on
the ethylene-propylene rubber showed a small effect, similar to that
obtained with the nitrile rubber using the above experimental procedure
and a larger effect when the procedure was modified to include alternate
long and short periods in contact. Figure 3 shows results obtained in
this way for 0.1h contacts alternating with longer (initially ca 20h)
contacts. The full line is that for the 'normal' dwell effect from
Figure 2, while the points show the intermittent 0.1h dwell results with
the peel rate plotted against the cumulative contact time. As can be

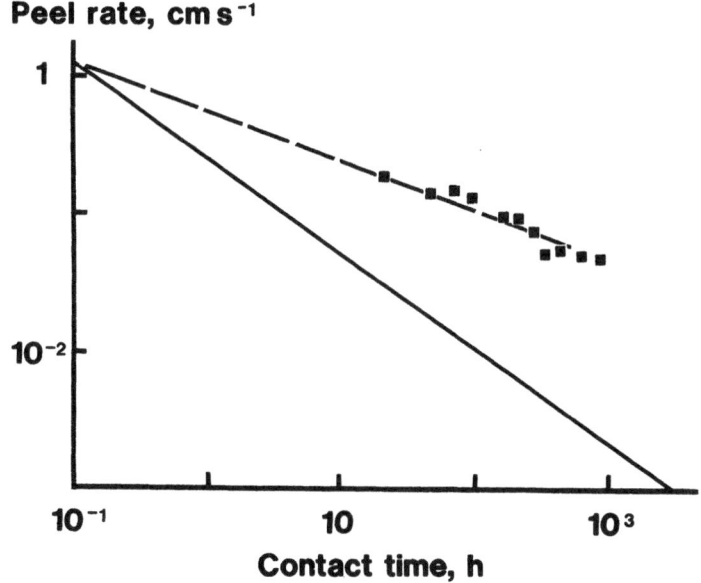

Figure 3. Dwell and memory effects for the ethylene-propylene rubber on
conducting glass: line for dwell results from Figure 2 ———; rate versus
cumulative contact time for 0.1h contacts alternating with longer
(initially ca. 20h) contacts —■— (peel energy = 8.5 J/m^2 throughout).

84

seen, the latter show a substantial decrease from the initial 0.1h dwell
result, the slope of the broken line being about half that for the normal
dwell results.

Thus some of the increased adhesion that occurs as a result of the
contact time is retained when a test-piece is detached from the substrate
and re-adhered. It appears that the test-piece has become 'conditioned'
in some way by the contact and is able to remember at least part of this
conditioning when it is re-adhered.

If a conditioned test-piece is stored out of contact with a
substrate the adhesion slowly declines until the initial level for an
unused test-piece is approached. Figure 4 shows the peel rate for the
same, fixed, peel energy as before plotted against time out of contact
(in air - care being taken to minimise access of dust to the test
surface), a short contact time of 2 minutes with the substrate being
allowed immediately before each measurement. As can be seen the rate
continues to increase over a period of 1000 hours or more, eventually
returning to the value obtained with an unused test-piece. Thus the
effect appears fully reversible. The loss of the conditioning or memory
can be greatly accelerated by immersing a test-piece in acetone.

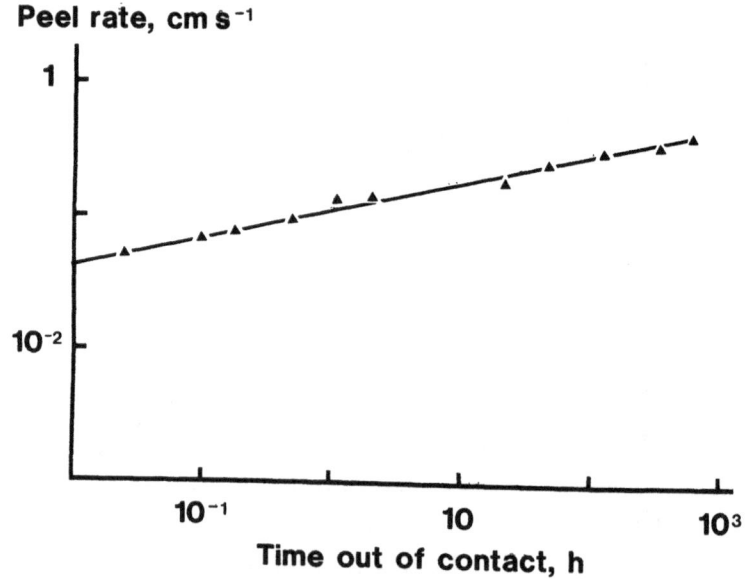

Figure 4. Decay of the memory for the ethylene-propylene rubber: peel
rate for intermittent 2 minute contacts with conducting glass against
cumulative time of storage in air (peel energy = 8.5 J/m^2).

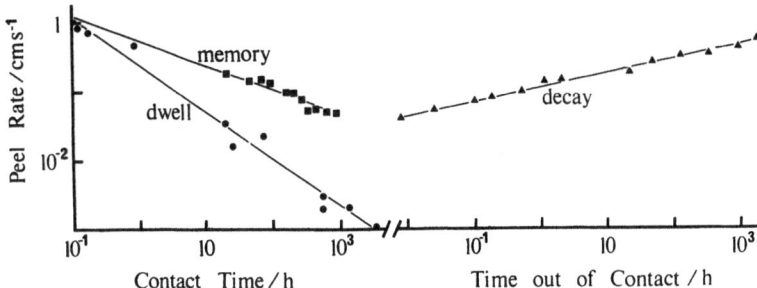

Figure 5. Summary of results of Figures 2-4 showing dwell, memory and decay effects for the ethylene-propylene rubber and conducting glass (peel energy = 8.5 J/m^2).

Figure 5 summarises the dwell, memory and decay observations. If a test-piece is taken through a long dwell and decay cycle and then re-adhered, the dwell and memory effects occur again in essentially identical fashion. Subsequently, the decay also occurs similarly. Thus the processes involved appear to be fully reversible and repeatable.

If a test-piece for which a memory effect has developed is re-adhered on a different area of glass, the improved adhesion is substantially retained: thus the memory appears mainly to be associated with the rubber.

Peel Characteristics

Hitherto, the variation in peel rate with time for a constant peel energy has been examined. If, instead, the variation in peel rate with peel energy for a given contact time is examined, curves such as that shown in Figure 6 are obtained. Similar time-dependent failure curves are observed for cohesive failure in non-crystallizing rubbers. Over a limited range such curves are often found to obey a relation of the form

$$r = BP^b \qquad (3)$$

where B and b are constants, i.e. a plot of log r against log P gives a straight line relationship (of slope b). This is the case for the upper part of the curve shown in Figure 6, the value of b being about 3, which is not dissimilar to the value obtained in cohesive failure with another non-crystallizing rubber—styrene-butadiene rubber [1]. At lower energies the simple power law relationship breaks down, the curve becoming steeper. Similar behaviour is again observed in the case of cohesive failure, the steeper slope indicating the approach (with decreasing energy) to a threshold energy below which mechanical failure

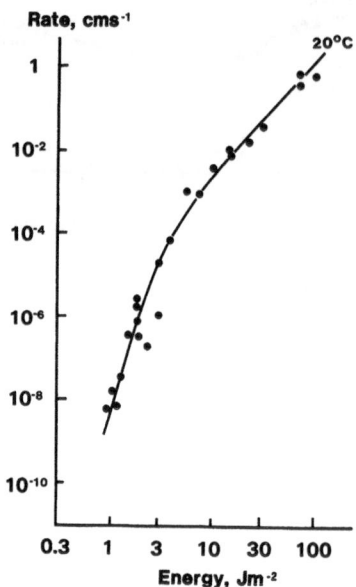

Figure 6. Peel rate versus peel energy for the ethylene-propylene rubber in contact with conducting glass at room temperature (contact time ca 1000h).

is initially absent [10]. Although the adhesion results in Figure 6 curve downwards, they give no clearcut indication of the existence of a threshold energy although from subsequent data it appears that a threshold does exist and that the lowest results are in fact very close to it. Before considering further data, the theoretical estimation of the threshold will be discussed.

THEORY FOR THE THRESHOLD ENERGY FOR ADHESION WHEN ONE OF THE ADHERENDS IS A HIGHLY-ELASTIC MATERIAL

In considering the theoretical calculation of the threshold energy for interfacial failure, it is convenient first of all to recall the corresponding calculation for the cohesive failure case for a highly-elastic material [11]. The basis of this is outlined below.

Cohesive Failure

If the situation at a crack tip in a highly-elastic material is considered, as illustrated in Figure 7, then ahead of the tip will be chains whose endpoints (crosslinks) lie on opposite sides of the plane of

87

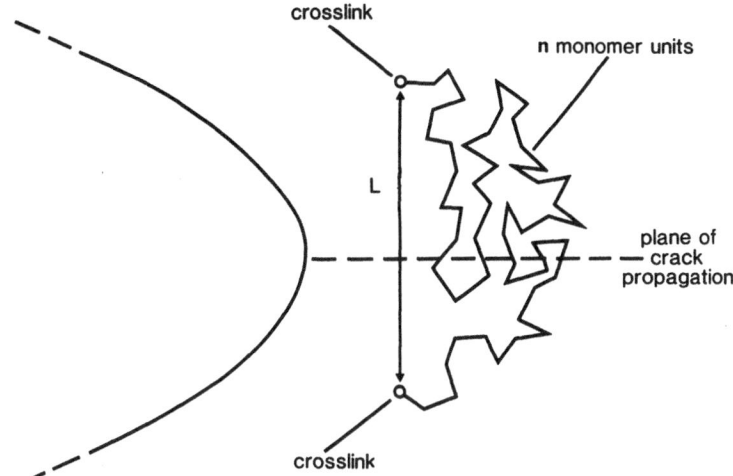

crosslink

n monomer units

L

plane of
crack
propagation

crosslink

Figure 7.   Schematic diagram illustrating a polymer chain crossing the
fracture plane ahead of a crack tip.

propagation. In order for the crack to propagate all such chains must be
broken. If the material is in the rubbery state, so that the chains are
freely vibrating, forces will be transmitted primarily via the
crosslinks. Prior to rupture, each chain, which in the unstrained state
is randomly coiled (as shown schematically in Figure 7), will first be
drawn essentially straight as the force increases and then bond
stretching and valence angle distortion will occur until finally one bond
breaks; at this point the force in all the other bonds between the
crosslinks will be essentially the same. Thus in order to break one bond
in a chain it is necessary to take each other similar bond in the chain up
to virtually its breaking point. Thus, if there are n monomer units in a
chain, each of which stores an energy U at the rupture force of its
weakest bond, then the energy required to rupture the chain will be nU.
If the number of chain vectors (L in Figure 7) crossing unit area in the
unstrained state is α, then the minimum fracture energy required may be
written as

$$G_o = \alpha nU \qquad (4)$$

where G_o is the minimum energy required to cause unit increase in the
area of one fracture surface referred to the unstrained state.

If the vector distance between the crosslinks at the ends of a chain
is L (c.f. Figure 7), then if the chain vectors are randomly oriented, it
can readily be shown that

$$\alpha = \frac{1}{2}LN \qquad (5)$$

where N is the number of chains per unit volume. N is equal to ρ/nm, where ρ is the density and m the mass of a monomer unit, and $L = \delta \ln^{\frac{1}{2}}$, where l is the length of a monomer unit and δ a factor reflecting the limited flexibility of the chains. Substitution in equation (5) gives

$$\alpha = \frac{1}{2} \frac{\delta \rho l}{m n^{\frac{1}{2}}} \tag{5a}$$

and then in equation (4) yields

$$G_o = \frac{1}{2} \frac{\delta \rho l U}{m} n^{\frac{1}{2}} \tag{6}$$

This calculation assumes the chains to be of uniform vector and contour lengths, but allowance for various distributions of these lengths only affects the value of the constant slightly; correction for wasted ends of the parent long-chain molecules would be expected to alter the dependence on n, particularly at low degrees of crosslinking [11].

Of the parameters in equation (6), ρ, l and m are basic parameters for a particular material, U can be estimated from known covalent bond strengths and stiffnesses, while estimates of n can be made from the statistical theory of rubber elasticity (for typical vulcanizates n is of the order of 100) and of δ from photoelasticity or other measurements. Substitution of representative values in equation (6) yields $P_o \sim 20J/m^2$, which compares with experimentally observed values of about $50J/m^2$ for various elastomers [10,11]. This appears satisfactory agreement in view of the neglect of additional energy losses and various other uncertainties in the theory. The substantial independence of elastomer type observed experimentally for the threshold energy contrasts with other strength properties but is as expected theoretically as the commoner elastomers all have carbon-carbon backbone chains.

Interfacial Failure

The situation for a vulcanized rubber adhering to a rigid substrate differs in three respects from the cohesive failure case:

   (i)   the chain configurations near an interface will be different
         from those for chains crossing a plane in the bulk;

   (ii) for van der Waals' bonds, the bond strength will be
         considerably less;

   (iii) because of the lower bond strength an effect of temperature
         might be expected.

Several aspects of the chain configurations are likely to influence the adhesion energy. Since the chains must be 'reflected' from a rigid surface the number of chains contacting unit area of such a surface cannot be more than about half the total number (ζ) of chain segments lying in unit area of a plane in the bulk, where

$$\zeta = \frac{\rho l}{m} \tag{7}$$

the symbols being as before. This will give an overestimate of the number of contacts since some of the contacts will be by different segments of the same chain. Alternatively, the number of chains in contact can be taken as the number of chain vectors (∝) crossing unit area of a plane in the bulk, given by equation 5(a). This is likely to give an underestimate of the number of contacts. In either case the area of contact, assuming all the available surface can be utilised, will be about twice the cross-sectional area of a single chain. Also, for chains contacting the surface at more than one point the segment of a chain nearest a crosslink may be stretched up to the bond breaking force and then partially relaxed several times as successive points of attachment to the substrate are broken. As a result the total energy required to detach the chain will be increased by some factor - X say. However estimates of the magnitude of this increase suggest that it is not large.

Taking into account the above factors the threshold energy of adhesion (P_0) can be placed between the limits

$$\frac{1}{2} \frac{X \rho \ell u}{m} \gamma n^{\frac{1}{2}} < P_0 < \frac{1}{2} \frac{X \rho \ell u}{m} n \qquad (8)$$

with a geometric mean value of

$$P_0 = \frac{1}{2} \frac{X \rho \ell u}{m} \gamma^{\frac{1}{2}} n^{\frac{3}{4}} \qquad (8a)$$

The degree of crosslinking and hence n can be estimated from elasticity measurements. According to the Mooney-Rivlin extension of the statistical theory of rubber elasticity, n is given by

$$n = \frac{\rho k T}{2 C_1 m} \qquad (9)$$

where k is Boltzmann's constant, T the absolute temperature and C_1 the first elastic constant in the Mooney-Rivlin form of the strain energy function, which can be determined from stress-strain measurements in simple extension [12].

With the present adhesion system the bonding at the interface is simply the result of the van der Waals' attractions between the molecules of the rubber and the substrate. The strength of these bonds is uncertain but is likely to be much less than that of the covalent bonds involved in cohesive failure. For two plane parallel surfaces in close proximity (closer than ca 100Å) the stress arising from the van der Waals' attractions is believed to vary as [13]

$$\sigma = \frac{A}{6 \pi D^3} \qquad (10)$$

where D is the distance between the surfaces and A the Hamaker constant. σ will have to be exceeded if the planes are to be separated. With S chain segments per unit area of surface the required force per chain

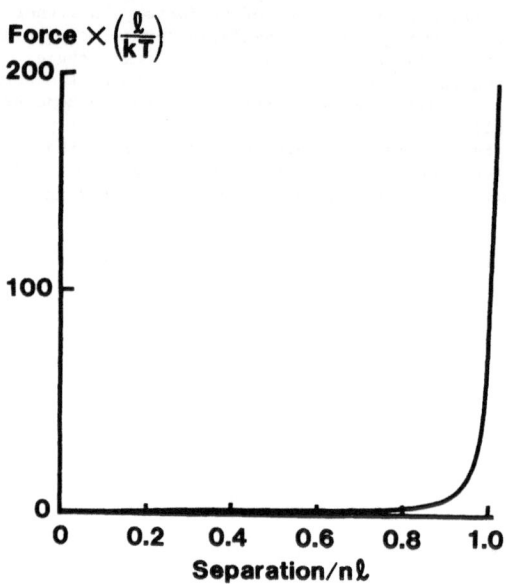

Figure 8. Force versus separation of the chain ends (L in Figure 7) for a single chain according to the inverse Langevin equation and in reduced variables form (notation is as in the text except that n and l here refer to statistical links rather than monomer units).

will be

$$f = \frac{2\sigma}{S} \qquad (11)$$

assuming each contact involves two segments, as discussed above. This force will have to be applied by tension in the rubber chains and as a result the chains will be extended and will store strain energy which will be lost when the contact is broken.

The nature of the force-displacement relation for a single chain according to the statistical theory of rubber elasticity is shown in Figure 8 for almost the whole range up to the fully extended length (nl). The curve has the form [12]

$$f = \frac{kT}{\gamma^2 l} \mathcal{L}^{-1} \frac{r}{nl} \qquad (12)$$

where r is the separation of the chain ends and $\mathcal{L}^{-1} \frac{r}{nl}$ represents the

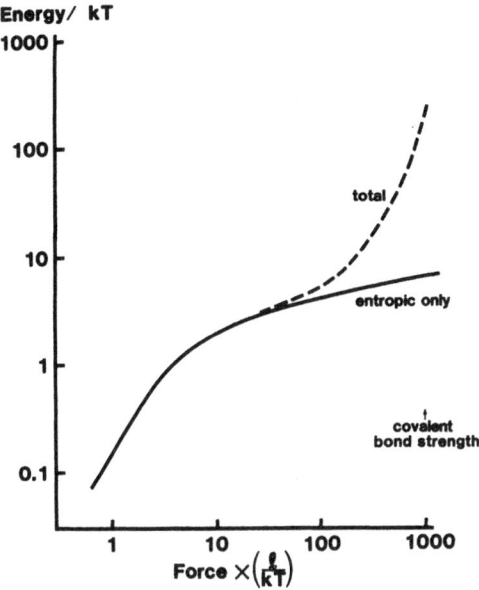

Figure 9. Energy stored per link versus force for a single chain (in reduced variables form).

inverse Langevin function given by β where

$$\mathcal{L}(\beta) = \coth\beta - \frac{1}{\beta} . \qquad (13)$$

Integration of equation (12) yields for the entropic contribution to the energy stored per monomer unit

$$U = \frac{kT}{\gamma^2}\left[\beta\coth\beta + \log_e \frac{\beta}{\sinh\beta} - 1\right] \qquad (14)$$

which for $\beta > \sim 3$ approximates to

$$U = \frac{kT}{\gamma^2}[\log_e 2\beta - 1] . \qquad (15)$$

Figure 9 shows the stored energy versus force relationship for a single chain plotted on logarithmic scales so that the dependence can be seen over a wide range. Because of the abrupt change in slope of the force vs separation curve in Figure 8 as the separation of the chain ends approaches the fully-extended value nl, the entropic contribution to the energy becomes very insensitive to the force at the higher forces shown in Figure 9.

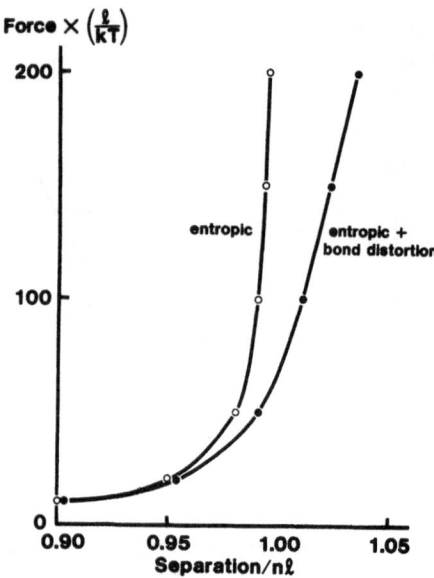

Figure 10. Force versus separation of the ends for a single chain, as in
Figure 8, but showing the effect of bond distortion at large extensions.

As the forces become large, so bond stretching and valence angle
distortion will allow appreciable additional deformation of a chain to
occur. The effect of this on the force vs separation relation is shown
in Figure 10 and on the energy vs force relation by broken line in Figure
9 which extends up to the covalent bond strength.

The possible effect of temperature on the adhesion threshold energy
appears difficult to predict, certainly on any simple basis. In the case
of cohesive failure the bond rupture energy is much greater than thermal
energies (cf Figure 9) and it might be expected on this basis that
temperature would have little effect. This has been found to be the case,
at least over a limited range. With the much weaker van der Waals'
bonds, the probable breaking energy is not much greater than kT, so that
the above argument does not apply. In order to clarify this situation
the effect of temperature on the adhesion has been investigated
experimentally.

In the present approach any kinetic effects in the failure are
ignored and the energy (per monomer unit) required for detachment is
taken to be that given by equation (15), which on substituting from
equations (7), (10) and (11) becomes

$$U = \frac{kT}{\gamma^2} \left[\log_e \left(\frac{2\gamma^2 Am}{3\pi D^3 \rho kT} \right) - 1 \right] \qquad (13)$$

which in conjunction with equation (8) or (8a) enables an estimate of P_o to be made.

NUMERICAL SUBSTITUTION AND COMPARISON WITH EXPERIMENT

The following values have been assumed for parameters involved in the theory:-

the multiple detachment factor, $\chi = 1$;

the density of the rubber, $\rho = 0.86 \text{gm cm}^{-3}$;

the length of a monomer unit, $l = 2.5 \times 10^{-8} \text{cm}$;

the molecular weight of a monomer unit, $m = 36 \text{amu} = 0.60 \times 10^{-22} \text{gm}$

[the average for the rubber used (47 mole % polyethylene, 53% polypropylene)].

Tensile stress-strain measurements on the rubber indicate a value for C_1 of 0.11 MPa, which on substitution into equation (9) yields $n = 260$. Estimates of γ are uncertain but a value of 2.0 has been obtained from photoelasticity measurements on a similar rubber [14].

A value has been taken for the Hamaker constant, A of 4×10^{-20}J; this is about the value quoted for the self-attraction of either polyethylene or polypropylene [15] and appears likely to be somewhat of an underestimate, as values for glass are probably higher. In view of the freely-vibrating long chain structure of the rubber molecules at the surface and the long time-scales for which the experimental results apply, it appears that the rubber molecules should be able to approach very close to the glass surface and a value for D of 3Å has been assumed.

Substitution of these values in equations (16) and (8a) gives a geometric mean value of

$$P_o = 0.37 \text{J/m}^2$$

with a variation of about a factor of 3 in either direction for the extreme limits.

The above values may be compared with the experimentally observed ones. Figure 11 shows results obtained at 50°C on the extracted ethylene-propylene rubber. At the higher end of the energy range the peel rates at 50°C are higher than those at 20°C (represented by the full line taken from Figure 6). This is as expected on the basis of the viscoelastic behaviour. At the lower energies the peel rates for the two temperatures become very similar, if anything those at 50°C being slightly below those at 20°C. The 50°C results extend to rates below

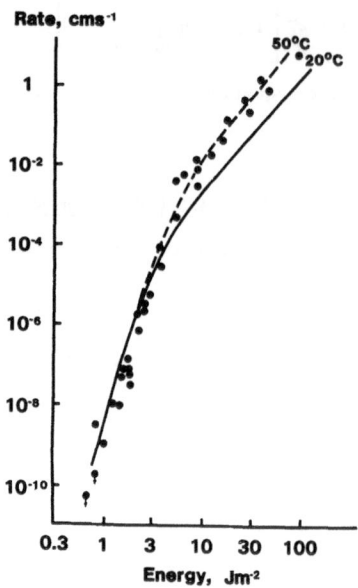

Figure 11. Peel rate versus peel energy for the ethylene-propylene rubber: against conducting glass at 20°C (from Figure 6, results against glass are very similar) ——— ; against glass at 50°C – – ● – –(contact times ca. 1000h).

10^{-10} cm/s and give indications of proximity to a threshold in this region. From the present data the threshold appears to be in the range 0.7–1.0 J/m^2 and seems likely to have little or no temperature dependence.

DISCUSSION

The observed P$_0$ values are two to three times the geometric mean of the calculated values. Bearing in mind the uncertainty of the latter, the neglect in the theory of energy losses except for those chains actually touching the surface (relaxation must also occur in adjacent chains in the body of the rubber), and the difficulty of the experimental measurements the agreement appears satisfactory. Indeed, a similar discrepancy is observed in the cohesive failure case (see earlier discussion and reference).

The substantial absence of an effect of temperature at the lower peel rates, albeit over a limited range, appears to justify the neglect of kinetic effects in the theory. At higher peel rates, the divergence of the results for different temperatures suggests the involvement of additional loss mechanisms (due to viscoelastic or other effects) as in the cohesive failure case.

The present observations are consistent with the existence of a threshold energy of adhesion, for a system involving a vulcanized elastomer, which is determined by the interfacial bond strength and the long-chain molecular structure of the elastomer. This threshold energy is much smaller than that required to produce cohesive failure in elastomers (ca 50 J/m^2), consistent with the relatively weak van der Waals' bonds for the adhesion system used, but greater than would be expected simply from surface or interfacial energies (ca 0.05 J/m^2).

Another investigation of the adhesion of vulcanized rubber to a rigid substrate is also reported in these proceedings [16]. In this investigation a dependence of the threshold energy of adhesion on the degree of crosslinking is inferred from failure measurements at relatively high rates. The observed dependence is close to that predicted by equation 8(a), viz $P_0 \propto n^{3/4}$ or (molecular weight between crosslinks)$^{3/4}$.

Although the present results are quite consistent with the existence of a threshold energy of the expected magnitude, further work is required to establish this on a sound basis comparable with that for cohesive failure. It is desirable that such work should include use of different rigid surface and vulcanized elastomer combinations, though the conflicting problems of blooming and oxidation present considerable difficulties with regard to choice of elastomer. Extension of the approach to different, more-strongly bonded systems would be a further development of interest.

The threshold energy observations are, inevitably, made at long contact times when the adhesion may be approaching a limiting value (although no evidence of this has as yet been obtained; continuing contact time effects would, of course, be expected to increase the measured P_0 values). The observations at shorter contact times of the dwell and memory effects also have implications from a mechanistic viewpoint. Thus whilst diffusion of impurity to form a surface bloom on a vulcanizate can undoubtedly affect the adhesion, the observations with the present highly-purified rubber (which shows no evidence of blooming even after very long periods) suggest that this is not essential for a dwell effect to occur. Similar conclusions apply to chemical reaction at the interface, in view of the inertness of the present material and the reversibility and repeatability of the dwell and memory effects. The memory observations also suggest that improved contact, due to dissolution of air bubbles trapped at the interface or macroscopic conformity of the surfaces as a result of creep and stress relaxation, cannot be the only mechanisms causing the dwell effect. Indeed, experiments in which the extent of surface contact was deliberately varied did show an effect but a relatively small one. A further striking illustration was provided by test-pieces which had been used for many experiments and had become highly-contaminated by dust adhering to the test surface; the contact with such a test piece could appear very poor but if there had previously been contact for a long period, the peel rate (at a given energy) could be orders of magnitude lower than that for a fresh test piece (although the latter might give apparently very good, essentially bubble-free contact).

While the mechanisms discussed above may each play some part in contributing to the dwell effect under particular circumstances it appears that another factor must also be involved, perhaps as the primary cause. There is some evidence of a change in the nature of the surface of the rigid substrate, perhaps due to dissolution of impurities by the rubber, but this is only small and the main effect appears to be associated with the surface of the rubber. It appears possible that rearrangement of molecules in the surface layer into more favourable configurations with respect to the substrate may be involved. The threshold energy calculation suggests that very close approach of the rubber to the substrate occurs. This is plausible in view of the flexible long-chain molecular structure of the rubber but it may be that initially many contacts are made that are not in the most effective orientation. The subsequent rearrangement of these contacts could involve breaking and re-making many of them, perhaps many times over, before the most favourable overall configuration is approached. It appears that such a cooperative mechanism might account for both dwell and memory effects and for the extraordinarily long time-scales involved, which are difficult to account for in terms of simple molecular movements in the rubber.

ACKNOWLEDGEMENT

The authors thank Dr A.D. Roberts whose suggestion stimulated the theoretical investigation.

REFERENCES

1. A fairly recent review of various aspects of fatigue and fracture of rubber is given in Prog. Rubb. Technol., 1983, 45, 89-143.

2. Lake, G.J. and Stevenson, A., J. Adhesion, 1981, 12, 13-22.

3. Lake, G.J. and Stevenson, A., in Adhesion 6, Applied Science, Barking, 1981, pp.41-52.

4. Counsell, P.J.C., in Aspects of Adhesion 7, Transcripta Books, London, 1973, pp.202-231.

5. Bates, R., J. Appl. Polym. Sci., 1976, 20, 2941-2954.

6. Roberts, A.D. and Othman, A.B., Wear, 1977, 42, 119-133.

7. Barquins, M., in Microscopic Aspects of Adhesion and Lubrication, Elsevier, Amsterdam, 1982, pp.369-387.

8. Lindley, P.B., J. Inst. Rubb. Ind., 1971, 5, 243-248.

9. Hata, T., Gamo, M. and Doi, Y., Kobunshi Kagaku, 1965, 22, 152-159.

10. Lake, G.J. and Lindley, P.B., J. Appl. Polym. Sci., 1965, 9, 1233-1251.

11. Lake, G.J. and Thomas, A.G., Proc. Roy. Soc., 1967, A300, 108-119.

12. Treloar, L.R.G., The Physics of Rubber Elasticity, Clarendon Press, Oxford, 3rd Edition, 1975.

13. See, for example: Tabor, D. and Winterton, R.H.S., Proc. Roy. Soc., 1969, A312, 435-450.

14. Furukawa, J., Yamashita, S., Kotari, T. and Kawashima, M., J. Appl. Polym. Sci., 1969, 13, 2527-2540.

15. Jacobasch, H.J., in Physiocochemical Aspects of Polymer Surfaces, Volume 2, Plenum Press, New York, 1983, p.641.

16. See Shanahan, M.E.R., Schreck, P. and Shultz, J: 'Role of Molecular Dissipation in Elastomer Adhesion'.

7

THE CURRENT SITUATION FOR HOLT-MELT ADHESIVES

R. T. Agger

Bostik Limited, Emhart Fastening Systems Group,
Leicester. LE4 6BW England

BS6138 describes a holt-melt adhesive as a thermoplastic adhesive that is applied in the molten state and forms a bond on cooling to a solid state. The adhesive usually melts between 100^{0}C, the lowest practical temperature to achieve a final bonded product to resist moderate warmth, to 200^{0}C where the application temperature, usually about 40^{0}C above this melting point, is reaching the decomposition point of the adhesive and probably the substrate as well.

Naturally occurring melts such as asphalts have been known for centuries and used for caulking boats and even possibly the 'slime' mentioned in the construction of the Tower of Babel. Candle and seal making extended the knowledge of waxes. The Industrial Revolution and more recently the great expansion of new raw materials produced by the petrochemical industry have given the adhesive formulations a wide range of suitable hot-melt ingredients. Hot-melt adhesives started to show significant growth inthe early 1050's which has accelerated to the present day often with growths of about 20% per year.

TABLE 1
Common Ingredients used in Hot-melts

Polyethylenes	Polyethylene vinyl acetate	Saturated Polyesters
Polyamides	Polyethylene vinyl acrylates	Polyurethanes
Poly vinyl acetates	Ethy cellulose	SBR (block type)
Atactic polypropylene	Butyl Rubber	Polyisoprene
Waxes	Plasticisers	Tackyfying resins
Inert Fillers	Antioxodants	

The general advantages of hot melt adhesives are outlined in Table 2.

TABLE 2

Advantages	Disadvantages
No solvent or water elimination	Remains thermoplastic
Good storage stability	May exhibit poor wetting
Readily adaptable to continuous	Need controlled heat - versatility
processing	loss
High coating thickness possible	Heat stability
with little shrinkage	Heat stability resistant adherents
	necessary
	Need specialised equipment for
	optimum use

At present the largest melt type is based on ethylene vinyl acetate resins with added resins, waxes, antioxidants and perhaps fillers and are used in such applications as general packaging, bookbinding and panel edge veneering.   The total hot-melt market is estimated to be approximately £21 million (1984).

Currently adhesives are becoming just part of a total substrates joining operation where the total cost of the end produce is the main object.   Machinery developments, improved factory methods, new market requirements and new raw materials are all interactive to change the hot-melt scene with time.   By way of illustration four examples are high-lighted:

1.   The use of EVA/resin type melts by the general public with better 'glue guns' (Fig. 1).   It is estimated that some third of a million households now own a glue gun.

2.   Competition in the shoe industry has been met with improved machinery, production lines and components.   Melt set-up times of the order of one second are usually required now.   These are provided by chemically made uncompounded polyesters and polyamides.

Fig. 2 Illustrates the various shoe parts requiring melts.

Fig. 3 Shows a hot melt lasting machine.

Fig. 4 Shows the various types of hot-melts required in shoe manufacture.

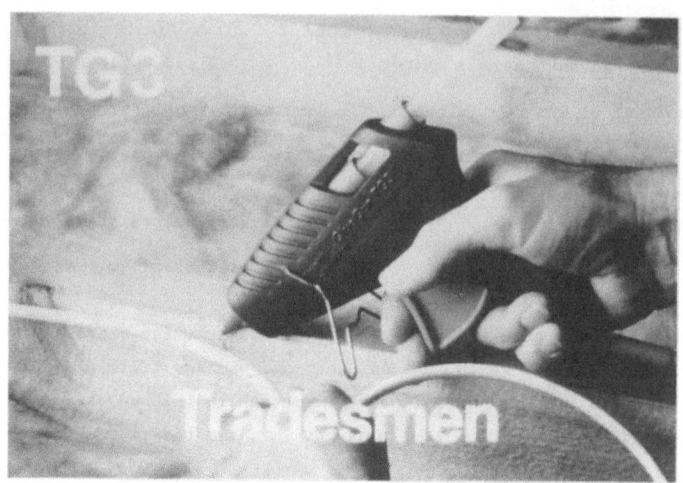

Figure 1. Glue Gun.

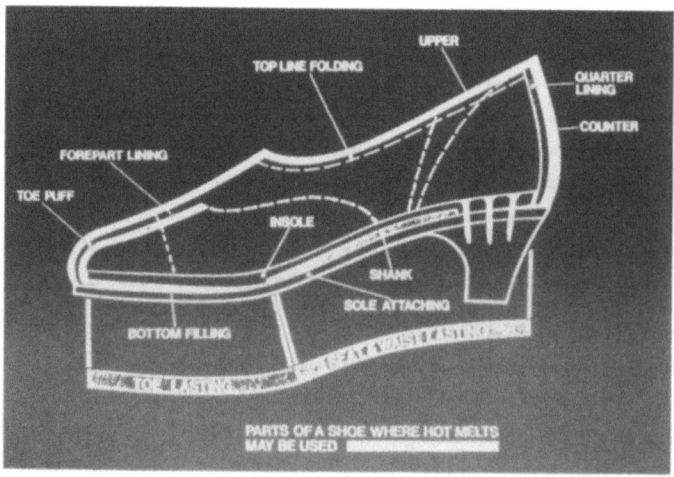

Figure 2. Various shoe parts requiring melts.

Figure 3. A Hot melt lasting machine.

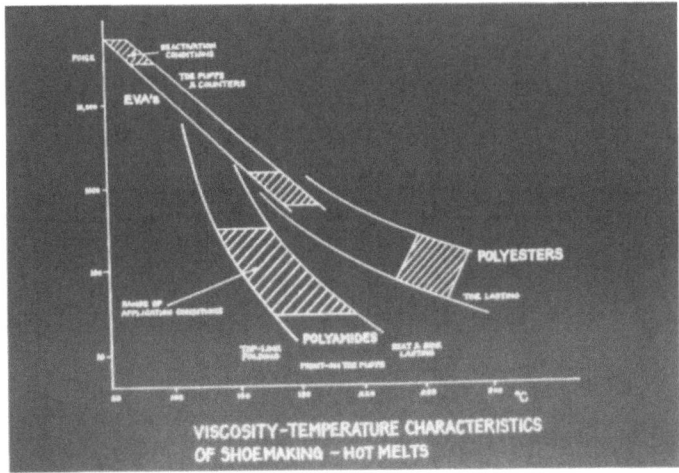

Figure 4. Various types of hot melts required in shoe manufacture.

Table 3 shows variation of saturated polyester melting points with
diol/diacid composition

TABLE 3
Saturated polyester m.p. v diol/diacid

Acid	Diol				
	2C	4C	6C	DEG	PG
Terephthalic	260	225	148	65	106
Isophthalic	A	140	75	55	80
Phthalic	63	17	0	10	45
Adipic	47	54	57	-25	-23
Azelaic	35	46	-	-	-

3. Energy conservation has meant a greatly expanding double glazed
window manufacturing industry.   An increasing percentage of such
windows produced in the UK are currently based on butyl rubber
based hot-melt sealants.

Fig. 5 shows the application of such a product where viscosities of
several thousand poise are being pumped at approximately 200°C.
Table 4 gives some of the important characteristics of this melt.
The main advantage of using a melt over two component polyurethane/
polysulphides is that the final units can be handled in about five
minutes.

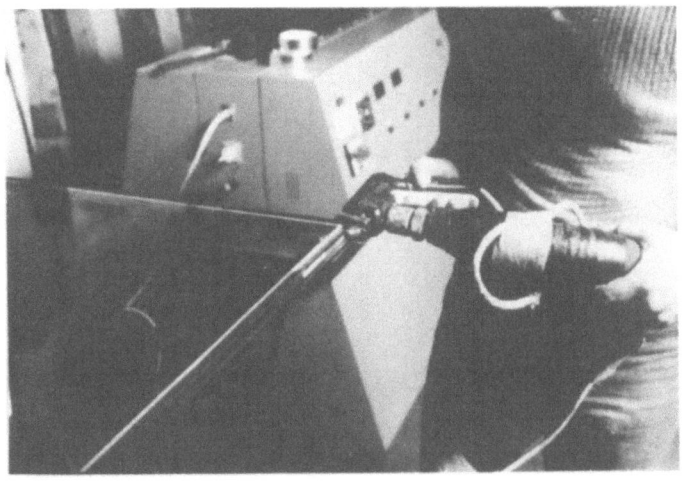

Figure 5. Application of a hot melt butyl sealant/adhesive.

TABLE 4
Hot-melt butyl sealant for double glazed unit manufacture -
general features

No pre-mixing - no waste

Fast setting - Units can be transported within 5 minutes

Excellent adhesion - To glass and all spacer bars including mill
finished and anodised aluminium, galvanised
steel and polycarbonates

Low moisture vapour   Typically 0.2g/m² /day (2mm sealant film)
transmission rate -

Typical viscosity -   400 Pas (4000 poise) at 180°C

Flexibility -   Satisfactory from -35°C to +70°C

Fogging of sealed   None
units -

4.   A disadvantage of hot-melts is that they have poor resistance to heat/
solvents.   One way of overcoming this is to use a moisture curing
PU melt.
This is basically a short M.W. urethane prepolymer tipped with
isocyanate groups.   These latter groups will react with water (from
the air and substrates) to cross-link via urea groups (Table 5).

TABLE 5
P.U. reactions

$$-NCO + -OH \longrightarrow \left[\begin{array}{c} O \\ \| \\ -N - C - O- \\ H \end{array} \right]$$

urethane

$$-NCO + H_2O \longrightarrow \left(\begin{array}{c} O \\ \| \\ -N - C - OH \\ H \end{array} \right) \longrightarrow -NH_2 + CO_2$$

$$-NH_2 + -NCO \longrightarrow \left[\begin{array}{c} O \\ \| \\ -N - C - N- \\ H \quad\quad H \end{array} \right]$$

urea

Another feature of this type of melt is its ability to crystallise rapidly. The the sequence of events from applying the melt at its application temperature (about 100°C) is:

a) Cooling: Physically in seconds going from about 150 poise liquid to an amorphous solid.

b) Crystallising: Physically going to a white, tough wax in about 2 minutes.

c) Reacting: Chemically/physically the -NCO/water reaction occurs leading to a greatly increased MW/toughness - typically 1mm cured in about 2 hours.

A specific example of its cost effective use is in the manufacture of TV cabinets where the total speed of box construction and the fact that the final adhesive is so tough that additional stiffening struts to keep the cabinet rigid are not required outweighs the initial high cost of the melt, and dispensing equipment.

8

CHEMISTRY OF REACTIVE URETHANE ACRYLATE ADHESIVES

DR. BOYD COORAY
Research and Development Manager
Baxenden Chemicals Limited
Paragon Works, Baxenden, Accrington, Lancashire,
BB5 2SL, England.

INTRODUCTION

Reactive urethane-acrylates of specific composition have been designed and developed as thermoplastic precursor macromolecules for yielding high performance thermoset structural adhesives with bond strength retention at temperatures up to 220°C. These products are members of a range entitled 'Xenacryl Reactive Adhesives' (XRA) and their development was directed at meeting the following requirements in a single component system:

(a) application temperature below 120°C and preferably
     at 60°-100°C,

(b) thermoplastic behaviour during application,

(c) zero volatile emission,

(d) non-hazardous during handling, storage and application,

(e) stability towards moisture and under storage,

(f) control over melt viscosity,

(g) control over rate of solidification,

(h) control over rate and extent of cross-link formation during post
     curing,

(i) choice of cross-linking conditions.

These criteria have to be examined in the light of existing structural adhesives and such an examination reveals that single component systems often fail to meet the requirements relating to stability, rate of crosslinking and/or environmental safety. Two component systems in usage generally yield a high level of properties but are limited by the hazards associated with their components, high curing temperatures and/or cost of materials and application equipment.

The chemistry of urethane-acrylate oligomers is well known and has been reviewed in a number of articles [1-4]. Urethane-acrylates are prepared by the interaction of diisocyanates with diols or triols or their admixtures to yield isocyanate terminated prepolymers which are then reacted with hydroxyacrylates or hydroxymethacrylates to produce terminal urethane-acrylate moeities as illustrated in Scheme 1.

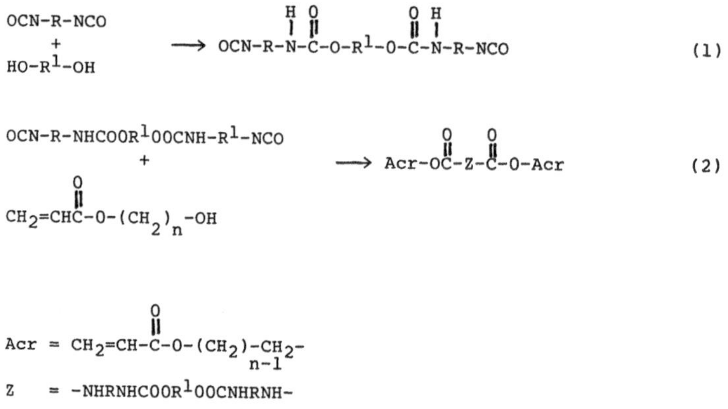

$$
\begin{array}{c}
\text{OCN-R-NCO} \\
+ \\
\text{HO-R}^1\text{-OH}
\end{array}
\longrightarrow
\text{OCN-R-N-C-O-R}^1\text{-O-C-N-R-NCO}
\qquad (1)
$$

$$
\begin{array}{c}
\text{OCN-R-NHCOOR}^1\text{OOCNH-R}^1\text{-NCO} \\
+ \\
\text{CH}_2\text{=CHC-O-(CH}_2)_n\text{-OH}
\end{array}
\longrightarrow
\text{Acr-OC-Z-C-O-Acr}
\qquad (2)
$$

$$
\text{Acr} = \text{CH}_2\text{=CH-C-O-(CH}_2)_{n-1}\text{-CH}_2\text{-}
$$

$$
\text{Z} = \text{-NHRNHCOOR}^1\text{OOCNHRNH-}
$$

Scheme 1. Urethane acrylate oligomer preparation.

Reaction (1) usually employs a slight molar excess of the di-isocyanate while Reaction (2) requires a slight molar excess of the hydroxy functional acrylate/methacrylate.

The molecular weight and molecular weight distribution of the product mixture generated by reaction sequences similar to that shown in Scheme 1 are controlled by adjusting the polyol/isocyanate ratio and by the choice of appropriate molecular weight polyols.

It is well known that in addition to the molar ratio of reactants employed, reaction conditions such as temperature, pressure, agitation, catalyst addition and relative reactivity of isocyanate groups and hydroxy functionalities determine the exact composition of the product mixture obtained in urethane acrylate formation.

The reactivity of isocyanate groups could also be modified by the introduction of protective groups or blocking agents that react with isocyanate groups at ambient temperatures at rates exceeding the reactivity of common polyols at elevated temperatures. Blocking agents such as 3,5 dimethyl pyrazole react with aliphatic isocyanates at room temperature thereby enabling control over molecular weight and composition during the reaction between diisocyanates and polyols.

$$\text{OCNRNCO} \quad + \quad \text{[3,5 Dimethyl pyrazole structure]} \quad + \quad \text{HOR}^1\text{OH}$$

3,5 Dimethyl pyrazole

$$\downarrow 15\,^\circ\text{C}$$

$$\text{OCNRN-C-N} \quad + \quad \text{HOR}^1\text{OH} \qquad\qquad (3)$$

Reaction of blocking agents with diisocyanates in the presence of polyols.

A significant advantage of employing blocking agents such as 3,5 dimethyl pyrazole is that the blocked isocyanate groups may be regenerated by heating to 120°-140°C or by the addition of diamines such as Jeffamine D2000 at 15°-60°C.

$$\text{Jeffamine D2000.} \qquad \text{H}_2\text{NCHCH}_2-[\text{OCH}_2\text{CH}]_x\text{NH}_2 \qquad x = 33 (\text{Average})$$

MATERIALS AND METHODS

The materials employed in the development of the XRA range of products were selected from 4,4' diphenyl methane diisocyanate, isophorone diisocyanate, commercial polyols and polyamines, 3,5 dimethylpyrazole, hydroxyalkyl acrylates and hydroxyalkylmethacrylates.

A brief survey of common raw materials employed in urethane acrylate oligomer synthesis illustrates relative differences in reactivity that contribute to the composition of the final product.

Commercial toluene diisocyanate is available as a pure 2,4 toluene diisocyanate or as a 80:20 mixture of the 2,4 and 2,6 isomers.

Toluene diisocyanate

The reaction of the TDI isomer mixture with polyols could be carried out at a suitable diisocyanate/polyol ratio to yield almost exclusive conversion of the 2,4 isomer to the corresponding urethane, thereby leaving the 2,6 isomer for subsequent conversion or for removal by vacuum distillation. The reactivity difference between the isomers is brought about by steric hindrance in the proximity of the isocyanate groups in the 2,6 isomer and these effects also contribute to the level of physical and mechanical properties observed in the urethane products.

A difference in reactivity of a lesser magnitude is observed with the two isocyanate groups of isophorone diisocyanate. The isocyanate group attached to the alicyclic ring is found to be less reactive than the primary isocyanate group.

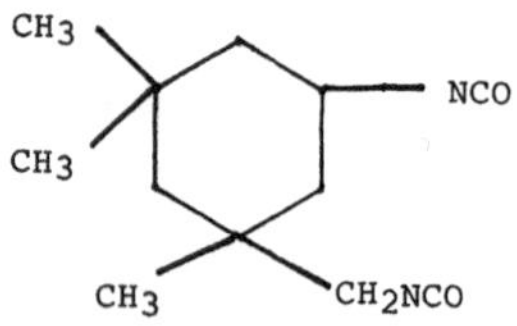

Isophorone diisocyanate

The methods of manufacture employed a combination of known and documented techniques that exploited the difference in reactivity between functional groups as outlined in the Introduction.

The products were analysed by a combination of techniques to establish levels of free isocyanates, free acrylic monomers, molecular weights and molecular weight distribution. Both gas liquid chromatography and head space analysis have confirmed that the products are free of isocyanates and their levels of acrylic monomers are below 0.1%.

RESULTS

The typical properties of the range of XRA resins are summarised in Table I.

TABLE I
Properties of XRA Resins

Polymer Type	Urethane Acrylate
Non Volatile Content	100%
Free Isocyanate Monomers	None
NCO Value	Zero
Free Acrylic Monomers	Less than 0.1%
Physical Form	Viscous liquid to hard solid
Melting Range	-5°C to 95°C
Glass Transition Temperature	-50°C to -30°C
Melt Colour	Colourless to pale straw
Melt Viscosity	200-6000 poise at 60°C
Melt Stability	5% increase in viscosity over 80h
Density	$0.95-1.2 \text{ kg.l}^{-1}$
Flash Point	>250°C
Peel Failure Temperature	180°C to 230°C
Available Form	Pellets, powder, film, viscous liquid.

The properties of this range of urethane acrylate resins enable environmental safety through the elimination of solvents and other volatiles such as monomeric isocyanates. The elimination of isocyanate moeities ensures that these products are unaffected by moisture during storage, handling, application and post-curing stages thereby overcoming the need for avoiding premature moisture ingress.

The melt viscosity of XRA resins is adjusted to suit a variety of applicational requirements by the choice of raw materials of suitable molecular weight and predetermined degrees of chain extension. The melt viscosity is further modified by the incorporation of high melt viscosity additives such as XRA compatible polymers.

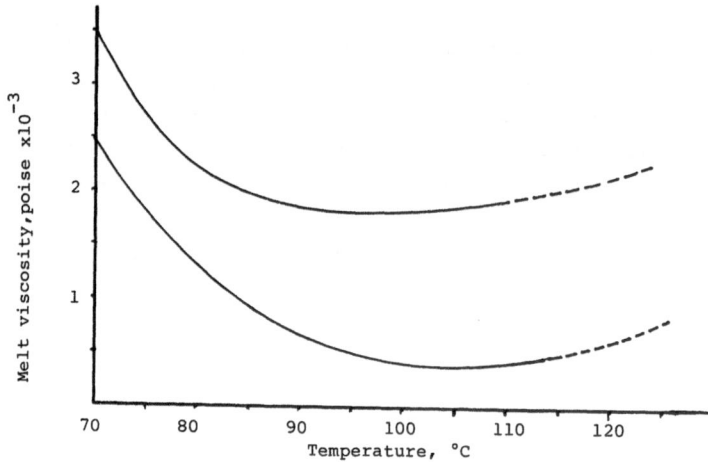

Figure 1. Effect of Temperature on XRA melt viscosity

As shown in Figure 1, both initial melt viscosity and its temperature dependence are a function of the type of XRA resin and its composition thereby enabling the design of adhesives to meet application requirements.

As temperature exceeds 120°C, the XRA resins exhibit an increase in viscosity due to the onset of cross-linking reactions which gradually convert the thermoplatic melt into a thermoset solid. The temperature and the rate of melt viscosity increase depend on the exact composition of a given XRA resin. At 60°-100°C, all XRA resins show excellent viscosity and colour stability enabling the melt to be held at this temperature range for over 80 hours with only a $\pm$ 5% change in viscosity and almost no measurable change in melt colour. Viscosity stability is further enhanced by blanketing the XRA melt with oxygen or air.

At 60°-100°C, the molten XRA may be applied to a variety of substrates using conventional hot melt equipment. Under these conditions there is no measurable chemical change as shown by molecular weight studies. The cooling of the melt upon contact with the substrate is dependent upon the temperature difference between the melt and the substrate as well as the composition of the XRA resin. Some XRA resins cool down to yield a tacky solid which crystallises over a period of hours to produce a tack-free mass. Others cool down to yield a tacky solid that undergoes rapid crystallisation to produce a tack-free mass in a few minutes.

A number of routes may be employed for converting XRA resins into thermoset adhesive bonds. This paper deals with results obtained under thermal curing conditions but separate studies have shown that XRA resins are readily cured by anaerobic and radiation techniques.

The cure rates achieved using thermal post-curing are illustrated in Figure 2.

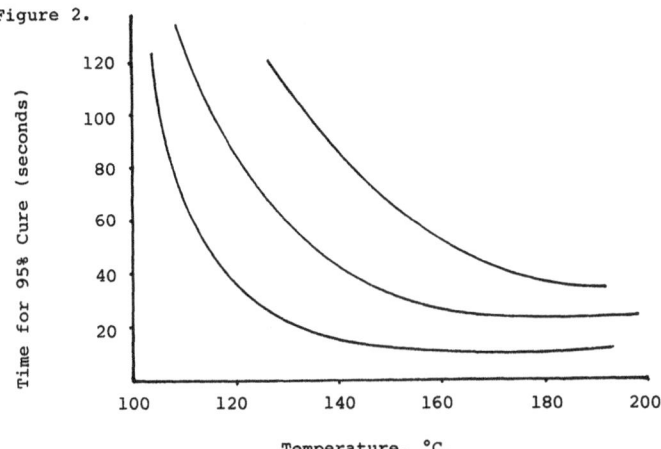

Figure 2. Cure Rates of XRA as a function of temperature.

In the temperature range 120°-210°C, XRA resins undergo chemical crosslinking reactions through carbon to carbon bond formation. The optimum temperature for curing a given XRA resin depends on its composition and on the combination of activators and initiators employed. Variations in resin composition and initiator design yields control over cure rates such that the resins may cure at either 140°C in 10 seconds, 140°C in 5 minutes or at 180°C in 10-300 seconds. The same XRA resin may be formulated to yield different rates of cure by careful control over the nature and level of activator/initiator systems. In general the fastest cure rates at the lowest temperatures are achieved with XRA resins that yield high cross-link density thermoset networks.

The results of curing XRA resins applied to substrates is largely to enhance adhesive bond strengths and their resistance to thermal, chemical and solvent exposure. These improvements together with changes occurring during the curing process are illustrated in Figure 3 which measures the thermal resistance of adhesive bonds as a function of temperature.

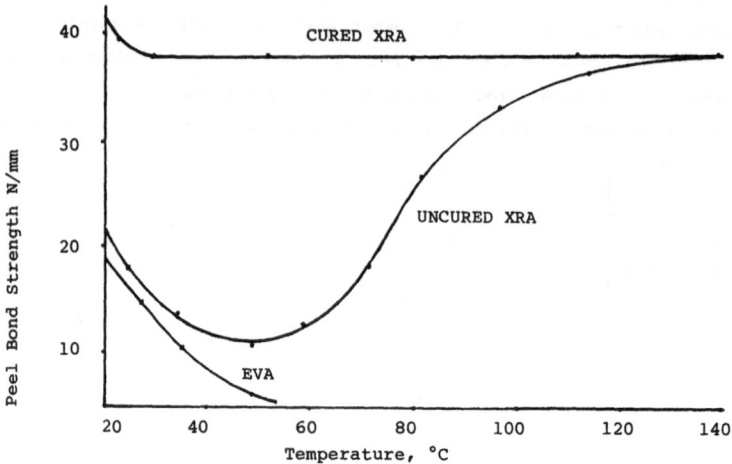

Figure 3. Bond strength of XRA as a function of temperature.

Figure 4 illustrates the level of control over the thermal curing process that takes place as a typical XRA resin is subject to heating above 100°C. It is evident that the cure profiles are relatively sharp and cure temperature for each resin could be predetermined by the selection of initiator/activator systems.

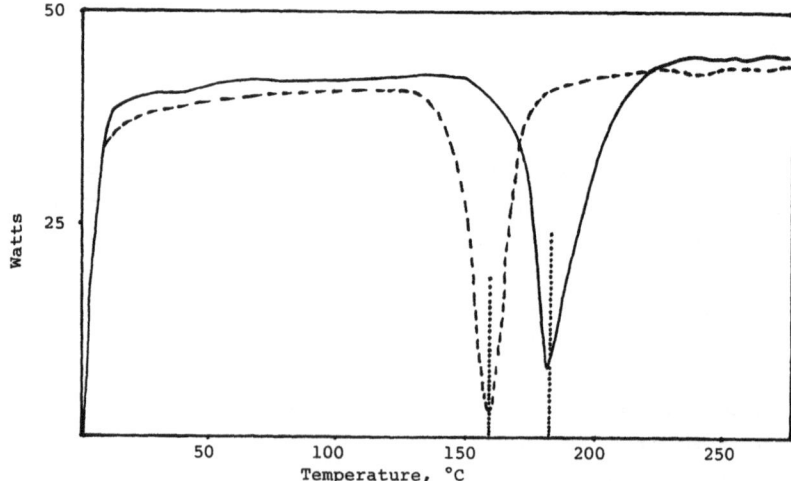

Figure 4. DSC scan for XRA Resins.

CONCLUSIONS

1. Reactive urethane acrylates of the XRA range are based on a versatile chemistry that enables modifications to be made to tailor adhesives for applications.

2. The XRA products exhibit a combination of desirable properties for meeting current and future demands for high performance adhesives.

3. Further chemical modifications are likely to lead to significant advances in XRA technology (5).

REFERENCES

1. C.G. Roffey: "Photopolymerization of Surface Coatings".
    Published by J. Wiley & Sons 1982.

2. C.B. Rybny: C.A. DeFazio, J.K. Shadidi, J.C. Trehellis, J.A. Vona
    Journal of Paint Technology 46(596) 1974 p.60.

3. Brittain: Journal of Applied Polymer Science 4 1960 p.207.

4. U.V. and E.B. Curing Formulation for Printing Inks, Coatings and Paints Edited by Dr. R. Holman 1984.

5. U.K. Patent Application Number 88.04686.

9

MOISTURE-CURABLE POLYURETHANE ADHESIVES AND SEALANTS FOR CIVIL ENGINEERING AND CONSTRUCTION APPLICATIONS

Janusz Kozakiewicz

Industrial Chemistry Research Institute

01-793 Warsaw, Poland

1. INTRODUCTION

The building industry is undoubtedly the major area of use of adhesives and sealants, consuming ca. 40% of their total production[1] and ca. 36% of special high peformance products (PU, silicones, polysulphides, acrylics, epoxies)[2]. This is because the product once tested and certified for civil engineering or construction applications is then widely and freely applied, sometimes in enormous quantities. Therefore research chemists involved in the adhesives business are attempting to develop commercial products that can meet severe building regulations and, together with civil engineers, are solving the complex problems that appear in bonding, coating and sealing of building materials. Some of these important problems concerning adhesion in civil engineering have already been presented quite a long time ago at the first Conference on Adhesion and Adhesives in this series[3]. During the 25 years that have passed from that Conference to the present one, a lot of entirely new products have been developed and used successfully in civil engineering and construction and knowledge of the adhesive properties of polymers and building materials has been distinctly improved.

This paper is devoted to moisture-curable polyurethanes that have recently become the very important materials for civil engineering and construction applications[4-7].

2. GENERAL REMARKS

There are limitations in the use of polymeric products as adhesives or sealants of building materials:
- they must be relatively cheap and easily available
- they must be very easy to apply in indoor and outdoor conditions
- their shelf-life and pot-life should be as long as possible
- they must adhere well to both porous (concrete, asbestos, ceramic, wood) and non-porous (glass, aluminium, rigid PVC) materials; priming of the surface is not usually possible
- they should preferably be self-extinguishing; or at least they should not contribute to the spread of flame
- they must exhibit excellent ageing resistance, since parts of building constructions are usually not exchangeable and must withstand many years of service.

116

One-package, moisture-curable polyurethanes are materials which, if properly formulated, are able to meet most of the above requirements. They are much cheaper than silicones, much easier to apply than two-component epoxies, they do not contain flammable solvents. They are 100% solids which is their advantage over aqueous dispersion-based products. Moreover, since the isocyanate group is extremely reactive, they can be easily tailor-made[4] to meet the different demands of the user.

3. ADHESIVES

The chemistry of moisture-curable PU adhesives used in the building industry is, in general, relatively simple; but it must be borne in mind that, with polyurethanes, even a minimum change in composition can affect the properties of the product to a great extent, and not all secrets of the manufacturers are revealed in the patents.

The major component of almost all adhesives of this type is the unblocked NCO-terminated PU prepolymer (Fig. 1a), usually with a polymer rather than a polyester backbone, due to better shelf-life attained by products synthetized from the former. Blocking of NCO groups which is sometimes used in sealants (Fig. 1b), is usually not recommended for adhesives since the presence of a blocking agent which is released during curing would result in decrease both in adhesion and cohesion of the bond.

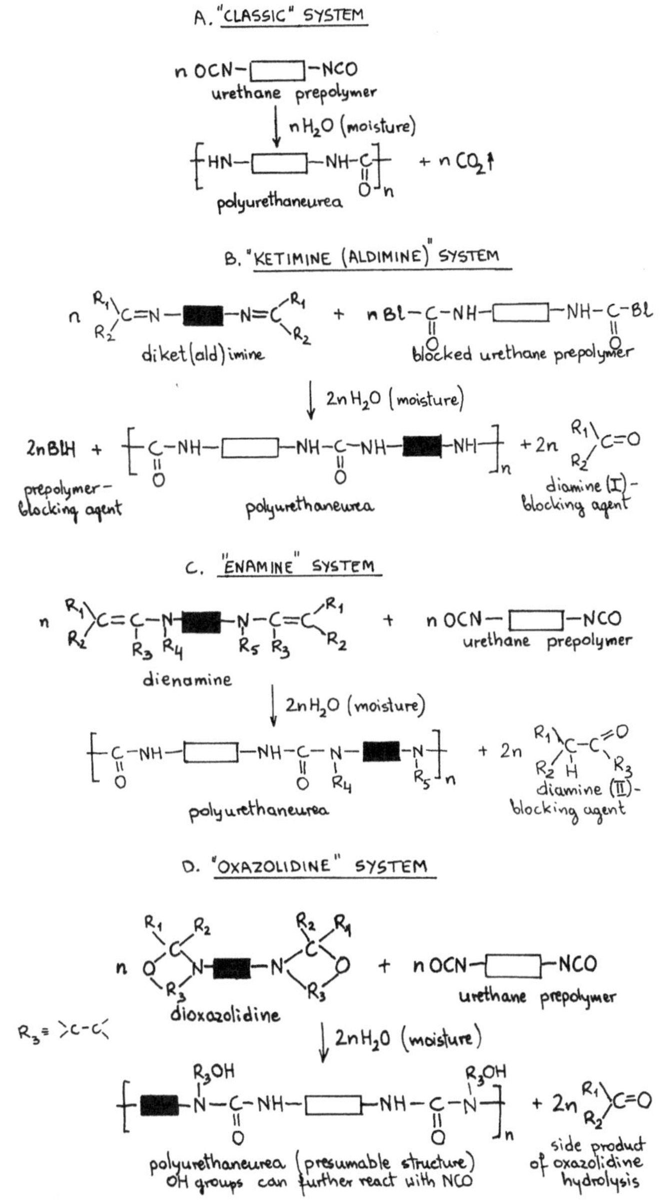

A. "CLASSIC" SYSTEM

n OCN—☐—NCO
urethane prepolymer

↓ nH₂O (moisture)

⟮HN—☐—NH-C⟯ₙ + n CO₂↑
                ‖
                O

polyurethaneurea

B. "KETIMINE (ALDIMINE)" SYSTEM

n R_1\C=N—■—N=C/R_1 + n Bl-C-NH—☐—NH-C-Bl
  R_2/   \R_2 ‖   ‖
                                O   O
diket(ald)imine   blocked urethane prepolymer

↓ 2nH₂O (moisture)

2nBlH + ⟮C-NH—☐—NH-C-NH—■—NH⟯ₙ + 2n R_1\C=O
        ‖   ‖ R_2/
        O   O
prepolymer-   polyurethaneurea   diamine (I)-
blocking agent   blocking agent

C. "ENAMINE" SYSTEM

n R_1\C=C-N—■—N-C=C/R_1 + n OCN—☐—NCO
  R_2/ | | \R_2 urethane prepolymer
         R_3 R_4 R_5 R_3
dienamine

↓ 2nH₂O (moisture)

⟮C-NH—☐—NH-C-N—■—N⟯ₙ + 2n R_1\C-C-C=O
 ‖   ‖ | | R_2/ H \R_3
 O   O R_4 R_5 diamine (II)-
polyurethaneurea   blocking agent

D. "OXAZOLIDINE" SYSTEM

R_3 = \C-C/

    R_1 R_2 R_2 R_1
n   \C/   \C/
   O N—■—N O   + n OCN—☐—NCO
      | | urethane prepolymer
      R_3 R_3
dioxazolidine

↓ 2nH₂O (moisture)

    R_3OH R_3OH
⟮■—N-C-NH—☐—NH-C-N—⟯ₙ + 2n R_1\C=O
     ‖   ‖ R_2/
     O   O
polyurethaneurea (presumable structure)   side product
OH groups can further react with NCO   of oxazolidine
                                           hydrolysis

Fig. 1   The potential moisture-curable PU formulations for
         building adhesives and sealants

         ☐ - urethane prepolymer backbone
         ■ - diamine backbone
         Bl - NCO--blocking group

In synthesis of NCO-terminated prepolymer technical grade MDI (Diphenylmethane-4, 4 - diisocyanate) is usually used because it is almost non-volatile and therefore less harmful to the user than other isocyanates. Free NCO content differs in the adhesives supplied by different manfacturers, but is usually not less than 3%, often reaching even 6-8%. For certain building materials (eg. concrete) the presence of NCO groups in the adhesive enables a chemical bond to be formed between adhesive layer and the adherend[8]. Even for the other building materials, the high polarity of NCO may play an important role in the adsorbing ability of the adhesive on the adherend surface.

It is obvious that the critical factor in the curing of this type of adhesive is the rate at which moisture can diffuse inside the bond; although practical results show that the strength develops quite fast (Fig. 2) even if the bonded materials are non-porous. The results shown have been obtained for small test specimens, but even for very large panels that are common in the building industry, the green strength obtained after 24 hours is high enough to handle or transport the bonded parts.

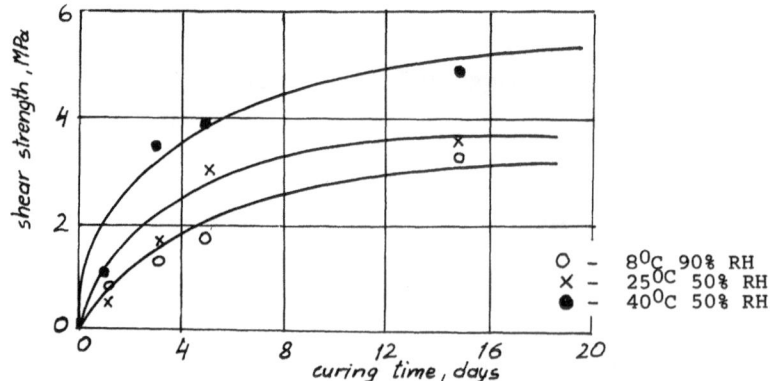

Fig. 2 Effect of curing conditions on shear strength of
Al-Al bonds made using moisture-curable PU adhesive

Twenty-four hours seems to be quite a long time, but in most building applications, shorter storage time is not required. It is claimed[9] that curing time can be cut tremendously (to 30 minutes) if a temperature of 60°-$75^{o}C$ and relative humidity 95% is applied to the sample (which means that the bonding must be done inside a closed steam chamber). The effect of air humidity and temperature on curing time of moisture-curable PU adhesives can be easily seen in Fig. 2 where the shear strength of the bond is plotted against time for samples cured under different conditions.

In the natural "wet spring" or "wet autumn" conditions $8^{o}C$, 90% RH, (they can be considered also as "English winter conditions"), the curing rate seems to be just the same as in "dry summer" or "winter indoors" conditions ($25^{o}C$, 50% RH), both being distinctly lower than that in "tropical conditions" ($40^{o}C$, 50% RH). What is more interesting, the bond strength in the latter conditions is higher.

A possible explanation is that at higher temperatures the viscosity of the adhesive layer is lower, so both the diffusion of moisture to the inside of the bond and evolution of CO_2 are therefore easier, which result in the bond being fully cured and bubble-free and hence results in higher bond strength.

An other explanation could be the easier formulation of Biuret crosslinks at higher temperatures in the reaction of NH groups of the polyurethaneurea with NCO groups of the prepolymer, which would certainly make the bond stronger. Certainly, the composition of the moisture-curable adhesive would affect the properties of the bond at least as much as it affects curing conditions, and is also responsible for the application properties of the product.

Since for most of the building applications the viscosity of the adhesive must be low to ensure good spreadibility, it is highly recommended that dilutents are used in moisture-curable PU adhesives. The problem could be, of course, easily overcome by applying conventional solvents (eg ketones or aromatic hydrocarbons), but not in the building industry, where toxicity and flammability requirements are very strict. The other disadvantage of using solvents is that some of the materials used in construction, eg polystyrene foam, are solvent sensitive.

It is also possible to synthetize polyurethane prepolymers used as adhesives with a very high excess of isocyanate, thus obtaining a low molecular weight and low viscosity product. This is, however, also not practical since in this case the excess of isocyanate must be distilled off (which is a very troublesome process) and the low-molecular weight prepolymer will not make a strong adhesive bond after moisture-curing. An interesting idea that is supposed to work well in practice is to use a low-molecular weight adduct of isocyanate with a polyhydric alcohol as a diluent[10].

Building adhesive must be easy to apply, but usually non-sagging properties are also required. This means thixotropy should be introduced to the commercial product. Using fumed silica not only adds thixotropy to the adhesive but also increases the bond strength due to the reaction of its active groups with free NCO. The other effect of fumed silica that is also reported is a reduction of the curing time of the adhesive[11].

Many other additives are used in moisture-curable PU to improve their properties, the most important are catalysts to shorten the curing time[12] and petroleum resins to improve tack[13]. Lack of tack and poor green strength can be considered as major disadvantages of moisture-curable PU adhesives. However, in certain applications, especially in the building industry, it is not too important a factor since the bonded parts may be stored for quite a long time before use.

Despite good prospects of using moisture-curable PU adhesives in construction and civil engineering[14 & 15], only a few references can be quoted that cover this subject. The most interesting area of application of these products seems to be the manufacturing of sandwich panels for lightweight structures or insulations[16]. The core of the panel is always polystyrene foam, while the outer parts of the sandwich can be either asbestos-cement boards or hardboards, particle boards, wooden boards, sheets of plywood or bitumen-impregnated roof papers depending on the application area. The sizes of the panels that are manufactured in the plant usually range from 0.5 x 1.0m to 1.5 x 3.0m.

A problem in bonding polystyrene foam is (see above) its sensitivity to organic solvents. Therefore, only aqueous dispersion or solventless products can be used. Dispersion adhesives form a non-crosslinked film which flows when subjected to load and higher temperature (a 70°C test is usually carried out), so only solventless, curable systems can be considered. An obvious solution seems to be to use amine-cured epoxies, but their disadvantages are: high toxicity of amine hardeners, the need for accurate proportioning of two components before application, and high viscosity and low spreadability of the resin-hardener mixture. Combinations of dispersed polyvinyl acetate and epoxy adhesives are also sometimes recommended, but they have very high viscosity and are also allergeric because of amine hardeners. Resorcinol-formaldehyde adhesives cannot be taken into account since they form very tough, but not elastic bonds. It appears then that moisture-curable polyurethane adhesives are ideal for this application.

They have practically unlimited pot-life, can be made as relatively low viscosity formulations and can produce a crosslinked bond of very good strength combined with elasticity and excellent ageing resistance. Despite the above mentioned advantages of moisture-curable PU adhesives, epoxies are predominantly used all over the world for sandwich panels, mostly because their price is lower. However, we found that moisture-curable PU can be applied at much lower quantities than epoxies; $0.2 - 0.3$ kg/m^2 as compared with 0.8kg/m^2 for epoxies, still ensuring very good properties of the panels.

The results of testing asbestos-cement/polystyrene foam sandwich panels manufactured using three different adhesives; moisture-curable PU, amine-cured elastified epoxy and resorcinol-formaldehyde showed[16] that though the load-deflection curves were very similar for all three materials, the greatest value of the maximum load to break the panel was obtained with the PU adhesive-bonded specimen. This result becomes obvious when the elasticities of the cured adhesive films are compared (see Table 1)[17].

TABLE 1

Adhesive	Temperature	Diameter of the Pin, mm			
		70	45	30	20
1	23	+	+	−	nt
	75	+	−	nt	nt
	X	−	nt	nt	nt
2	23	+	+	+	−
	75	+	+	+	−
	X	+	−	nt	nt
3	23	+	+	+	+
	75	+	+	+	+
	X	+	+	+	+

Elasticity of the Adhesive Films at Various Temperatures.

1 - epoxide adhesive with amine hardener
2 - elastified epoxide adhesive with amine hardener
3 - moisture-curable PU adhesive

+ - no cracks
− - cracked
nt - not tested
X - 50 cycles (-30 and +40°C)

It is noteworthy that the PU adhesive shows superior performance not only at room temperature and at 75°C but especially in thermal cycling which means excellent ageing resistance. It has been proved by the results of ageing sandwich panels in the water immersion/heating/drying/cooling cycles, and by the several years of service.

In Poland, asbestos-cement/polystyrene foam sandwich panels, made using moisture-curable PU adhesive, have been manufactured successfully for more than five years without any problems.

The second important application of moisture-curable PU adhesives in construction is bonding wood structures for buildings and parts of doors and windows. The advantages of using these adhesives instead of resorcinol-formaldehyde resins are those already mentioned above, together with the fact that they can be applied to wood which is not fully-dried and they do not attack wood tissue, as their pH is neutral. Since PU adhesives can be applied on one surface only and they do not contain fillers, the bond thickness is much lower than for resorcinol-formaldehyde resins (0.02 - 0.04mm as compared to 0.1mm). The results of severe tests made with moisture-curable PU bonded samples of pine-wood (eg 6 hours immersion in boiling water)[18] showed that in most cases the bond failure occurred in the wood and the bond strength was approximately the same as that obtained for resorcinol-formaldehyde adhesives. The advantage of PU adhesive over resorcinol-formaldehyde resin was more clear when the specimens were stored at 55°C for 24 hours and tested at the same temperature. Accelerated and real-time ageing studies proved the possibility of using moisture-curable PU adhesives for bonding of wood structures[19].

Footnote:

(a) The chemistry of moisture-curable hot-melts differs from that of PU adhesives described earlier in this paper. They are NCO-terminated thermoplastic elastomers obtained from polyesters of a high degree of crystallinity and diisocyanates. The initial strength of the bond is the effect of crystallization and the final strength is reached via moisture-curing.

Moisture-curable PU hot-melts[a] can also be considered as potential construction adhesives for wood structures since their bond strength is very high ($12N/mm^2$ has been reported[20]) and their advantage lies in very fast bonding. However, it appears that, at least for presently available formulations, steam chambers are necessary to complete moisture curing.

Another interesting area of possible application of moisture-curable PU adhesives is bonding of building or pipe heat insulation made of polyurethane foam. For this purpose, the adhesive should have non-sagging properties which can be easily obtained by modification with fumed silica.

Results of the investigations carried out in the Industrial Chemistry Institute in Warsaw showed that even after water immersion or thermal cycling of PU foam, the bond did not fail when the specimens bonded with moisture-curable PU adhesive were tested. This adhesive is now being used in civil engineering for bonding PU-foam insulation sheathing to iron pipes in hot water supply systems and is being tested by the Institute of Building Technology as the glue for sheathing of concrete buildings with PU foam.

Sealants

Unlike the adhesives, moisture-curable PU sealants are commonly used in civil engineering and construction applications together with silicones and polysulfides.

Their chemistry is sometimes much more complex than that described in this paper for the adhesives. Of course, the simplest formulations can be up to 50-60% filled, NOC terminated urethane prepolymers based on polyether polyols of M_W = 2000 - 6000 and diisocyanates, aliphatic isocyanates (eg IPDI - isophorone-diisocyanate) being especially recommended since they assure good weatherability of the product. Though this simple NCO terminated prepolymer system is still widely used in sealants[21], several novel formulations have been developed during the last few years[22]. The problem with sealants is that they contain a high percentage of fillers which makes their shelf-life shorter and thick layers must be cured with moisture which makes their curing time longer than for the adhesives.

Therefore, the direction of development here is quite clear: to achieve a product of long shelf-life and short curing time. Actually, two routes can be considered to reach this aim:

- blocking -NCO groups in the prepolymer and addition of a blocked curing agent that can be released rapidly in the reaction with moisture (Fig. 1b)
- leaving -NCO groups in the prepolymer unblocked and addition of a blocked curing agent that can be released not too rapidly in the reaction with moisture (Fig. 1c & 1d)

Despite the number of patents claiming formulations consisting of blocked prepolymers and blocked curing agents (diamines[23-30]), it seems that the second route has so far been mainly achieved in the commercial sealants used in the building industry.

127

The main reason is that, so far, it has not been possible to find NCO-blocking agents that would ensure fast deblocking by the amine at room temperature and at the same time would not be harmful and would not spoil the properties of the cured sealant. The other reason is the high sensitivity of blocked primary diamines (or diketimines) to hydrolysis which can make the shelf-life of the product very short if the package is not sealed adequately. Although in one of the German patents[31], an aldimine of specific structure is claimed to be safe in the mixture with the unblocked prepolymer, the common practice seems to be rather directed to using enamines [32 & 33] (Fig. 1c) or oxazolidines[22, 33-35], (Fig. 1d), which are less moisture sensitive. Certainly, also "mixed" (eg enamine-oxazolidine or aldimine-enamine) systems are possible.

It is noteworthy, however, that in the formulations where only the amine component is blocked, the blocking agent released during curing would remain in the cured product if it is not volatile enough. Furthermore, the structure of the cured product would be different from that obtained in direct reaction of NCO-Terminated prepolymer with moisture (compare Fig. 1a with 1b, c, d) which may result in different properties of them both after curing. It should be specially emphasized because the presence of low-molecular weight blocking agent and the different structure of the cured product must affect its ageing resistance.

On the other hand, all systems with a blocked curing agent are much less sensitive to curing conditions which is their great advantage, especially when they are applied outdoors. Here, again "enamine" or "oxazolidine" systems look much more reasonable than "aldimine" or "ketimine" systems with blocked prepolymer because the latter require the prepolymer-deblocking reaction to proceed and this process is very sensitive to temperature.

Once the substantial formulation of the sealant is fixed, the other additives can be chosen. The effects of filler and plasticizer on the properties of moisture-curable building sealants based on NCO-terminated prepolymer with a polyoxypropylene glycol backbone was carefully studied[36, 37] and it was found that the plasticizer/filler ratio was the most important factor affecting the rheological properties of the uncured sealant (See Fig. 3) while the mechanical properties of the cured product depend mainly on the plasticizer content.

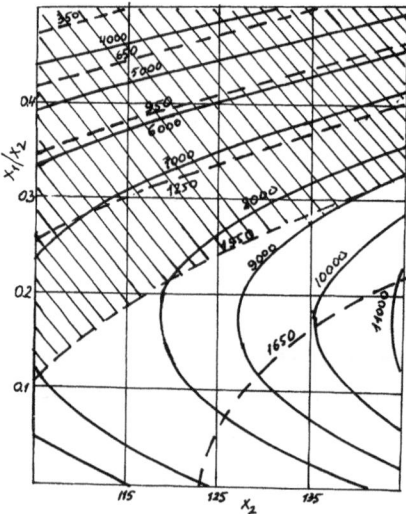

Fig. 3 Effect of plasticizer (filler X_1/X_2) ratio on rheological properties of the moisture-cured PU building sealant based on NCO-terminated prepolymer.

     — - low shear rate viscosity ($10^{-1}s^{-1}$)mPa's
     -- - high shear rate viscosity ($1s^{-1}$) mPa's
          The area of good spreadibility is hatched.

The thixotropy of this particular sealant was achieved by addition of fumed silica, which is the most common practice. But in certain formulations reaction products of diisocyanates and diamines or polyamines[22], fatty acid - polyoxypropylene glycol esters[38] or even PVC powder[22] are recommended as thixotropes. Other important additives to moisture-curable PU sealants are silanes (for better adhesion)[39, 40] and storage stabilizers (inhibitors of the reaction between NCO and water) eg benzoic acid[31] or benzoyl chloride[41].

Some commercial formulations used in building industry also contain extenders; especially favoured are bitumens, petroleum asphalt[42 - 44], coal tar[45] or coal tar pitch[46, 47]. Including them in moisture-curable systems is not simple because of compatibility problems and the possibility of uncontrolled reaction between active groups of bitumen components and NCO groups from the urethane prepolymer[45, 48, 49]. These problems can be solved by first preparing their isocyanates and mixing such modified extenders with the PU components. The extensive studies on compatibility, rheology and application properties of the blends of PU with asphalt - isocyanates[44] or coal tar pitch-isocyanates[46, 47] (CTPA adducts), showed that the optimum rheological and mechanical properties occurred at certain concentrations of the extenders in the blend. The specific internal structure of the blends is supposed to be responsible for this phenomenon.

130

TABLE 2

	1	2
Cone Penetration, °Pen	242	270
Sagging	none	none
Tensile Strength, MPa	2.1	1.1
Elongation at Break, %	128	102
Shore A Hardness, °Sh	55	70
Tack Free Time, Hours	<12	<12

Properties of Two PU/CTPA Blends Tested in the Building Industry

1 - ca. 40% CTPA
2 - ca. 65% CTPA

The results presented in Table 2 for two commercial formulations developed in the Industriual Chemistry Research Institute in Warsaw proved that even at very high content of CTPA in the blend with PU, the mechanical properties of the blend still remained quite good. Results of testing the blend containing ca. 65% CTPA as insulating roofing material for the building industry are presented in Table 3.

Both the fundamental moisture-curing system on which the sealant is based and the other additives affect not only the final properties of the cured material but also its curing characteristics[50, 51]. As has been already pointed out in this paper, this effect is very important, especially in building applications where curing conditions can vary from cold and wet to hot and dry depending upon the season and place.

Results with moisture curing of PU sealant based on NCO-terminated prepolymer presented in Fig. 4 and 5[53] show that neither too high nor too low a temperature or humidity is recommended for this particular product.

TABLE 3

No	Test	Before Ageing	After ageing (20 cycles)[a]
		Result	
1	Flow on vertical surfaces (45^0 angle, 5hrs, 70^0C) From concrete From roofing paper	no flow no flow	no flow no flow
2	Low temp flexibility -5^0C -20^0C	no cracks on 05 no cracks on 05	no cracks on 05 no cracks on 020[b]
3	Flow of the roofing paper bonded to concrete with PU/CTPA blend 45^0 angle 5 hrs, 70^0C	no flow	no flow
4	Bonding strength (for $20cm^2$ area) - roof pap+roof pap - roof pap+concrete	roofing paper failed in the other place than the bond	

a	- one ageing cycle:	1. uv irradiation	- 2 hrs
		2. rain 20^0C	- 2 hrs
		3. heating 70^0C	- 4 hrs
		4. room temp	- 16 hrs
		5. rain 20^0C	- 2 hrs
		6. freezing -25^0C	- 5 hrs
		7. room temp	- 17 hrs
		TOTAL:	48 hours

b - after 15 cycles - no cracks on 015

Results of testing moisture-curable PU/CTPA blend/CTPA content ca. 6.5% as insulating roofing material for building industry.

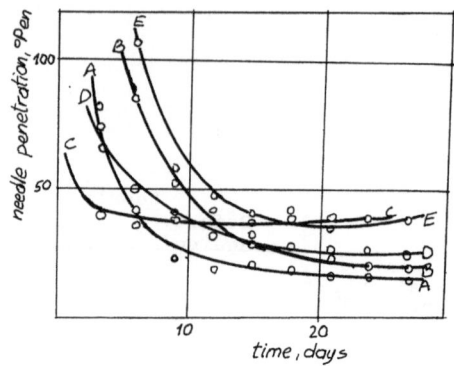

Fig. 4   Effect of temperature and hundityy on curing rate of
moisture curable PU building sealant based on NCO
terminated prepolymer

        —   25^{0}C 50% RH
        —   45^{0}C 10% RH
        —   45^{0}C 90% RH
        —   5^{0}C 90% RH
        —   5^{0}C 10% RH

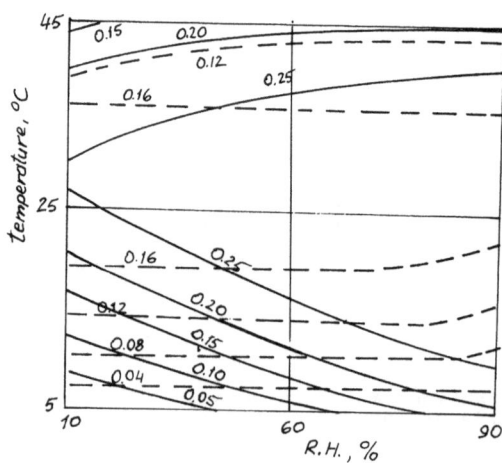

Fig. 5   Effect of temperature and humidity on tensile
strength (MPa) and 100% modulus (MPa) of moisture
cured PU building sealant based on NCO terminated
prepolymer.

133

This is probably the reason that the properties and curing characteristics of many commercial building sealants are given in technical data sheets, but only for the optimum conditions (usually for 20 - 25°C and 50 - 60% RH). However, it seems that the curing rate of certain "faster" systems, eg enamine containing forumlations, would be less sensitive to curing conditions that that of "slower" systems based on NCO-terminated prepolymers alone. The results presented in Fig. 6 prove that for a reasonable temperature range (over +10°C) the tack-free time for an enamine system based sealant remains almost constant with temperature.

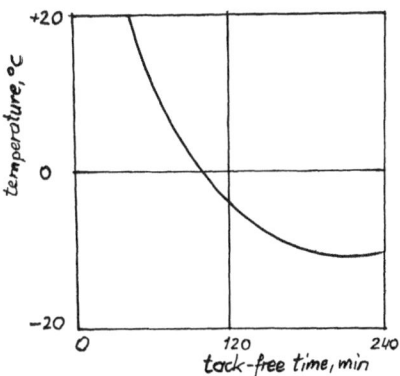

Fig. 6 Effect of temperature on curing rate of moisture curable sealant based on "enamine system"[33].

This is undoubtedly the great advantage of this type of sealant over traditional products based on NCO-terminated prepolymers, though the mechanical properties of the latter can sometimes be better (compare the results from Table 4 and 5). Moreover, it is not possible to predict the performance of the new formulation in outdoor building applications without the results of accelerated ageing tests. Since both the composition and the curing system can affect not only

134

TABLE 4

100% Modulus, MPa	0.22
Elongation at Break, %	450
Recovery After 100% Elongation, %	80
Shore A Harness, °Sh	15
Curing Rate at 20°C, mm/day	3
Storage Stability, months	>6
Sagging	none

Properties of Moisture-Curable Sealant Based on "Enamine System"[33]

TABLE 5

Results of accelerated ageing test (after 1000 hours) made for two different formulations of polyurethane building sealants.

1 = moisture-curable PU sealant based on NCO-terminated prepolymer

2 = 1 modified with storage-stabilizer (benzoyl chloride)

Conditions of test: UV lamp 4500W, RH = 60-70%
t = 30-35°C/70°C on the sample surface

Cycle: rain - 5 min, dry period - 25 min
B - before test   A - after test

	1		2	
	B	A	B	A
Tensile strength MPa	1.7-2.2	1.4-1.8	1.6-2.1	1.3-1.8
100% Modulus MPa	0.6-0.9	0.6-0.8	0.6-0.8	0.5-0.7
Elongation at break%	500-700	500-700	500-650	450-600
Shore A hardness °Sh	35-40	40-45	35-45	40-48

physical-chemical properties of the sealant but also its weatherability, accelerated ageing tests must be made if a new product is developed or any change is made in the old formulation. As it can be seen from Table 5, the addition of storage-stabilizer did not decrease the excellent weatherability of the NCO-terminated prepolymer-based sealant developed in the Industrial Chemistry Research Institute in Warsaw. These results were supported by the performance of the products in special severe ageing and durability tests (similar to ASTM C - 920) made in the Institute of Building Technology in Warsaw[52].

Presently, investigations are being carried out to change this good, but "slow", system into a "fast" one. In the industrial applications when very fast curing is necessary, one of the possible solutions is to use an amine curing agent as the second component of the NCO-terminated prepolymer-based sealant. Such a system with aniline-cresol-formaldehyde resin as curing agent has been recently developed in the Industrial Chemistry Research Institute in Warsaw[52, 54]. The recommended amount of curing agent is not stoichiometric, so only initial curing proceeds via $NCO-MH_2$ reaction and final curing is achieved in the reaction of NCO with moisture. Very good behaviour of this system in the curing process[55] and the excellent properties (especially weatherability), of the cured product, made it possible to use it as a structural sealant for manufacturing of insulated glass window panels. In this particular application, the fact that the NCO-terminated component is itself able to be moisture-cured, is the great advantage of this system over the others, especially polysulfides.

The reason is that uncontrolled change in the sealant/curing agent ratio sometimes happens during manufacturing. If this PU system is used all seals will be perfectly cured even in this case while the use of the other systems will result in damage of the uncontrolled production lot.

This is a good example of a general rule that any particular building application may require some special features of the sealant. Eg construction of chemical works would require a chemically resistant product, and construction of nuclear stations a radiation resistant one. It is nice to report that, according to the preliminary results, moisture-curable PU sealants perform well even in the latter application.

Generally speaking, the number of possible building applications of moisture-curable PU sealants is enormous, since they can be used in all joints of concrete, glass, wood, aluminium, steel or iron, stone, ceramics or porcelain, PVC and many other plastics, especially when they are subjected to expansion and contraction movements. In Table 6, some applications of moisture-curable PU sealants in the building industry are presented and in Fig. 7 some examples of the joints are shown.

To conclude this review of moisture-curable polyurethane adhesives and sealants for civil engineering and construction applications, it seems reasonable to point out again the great potential for development of this sort of material. The high reactivity of NCO-group is still the source of new

137

TABLE 6

Filling of vertical or horizontal joints between concrete panels in the main structure of the building

Sealing of gaps between window or door wooden frames and the concrete of brick walls of the building

Sealing of vertical or horizontal joints of lightweight structures in buildings (eg curtain walls) bathroom fixtures

Sealing of all joints and gaps in flashing and roofwork[b]

Sealing of all joints in waterproofing insulations made with PVC film (eg terraces, balconies, swimming pools)

Sealing of all joints in air-conditioning and ventilating systems in buildings

Sealing of all concrete and PVC pipe joints in sewerage systems[c]

Sealing of steel (iron) and PVC pipe joints in drinking water supply systems

Sealing of leakages in pipe joints in city gas supply systems[c]

Sealing of butt joints of concrete panels in underground tunnels[b]

Structural glazing, especially in manufacturing of insulated glass window panels[a]

a — two-component systems are preferred
b — polyurethane-bitumen formulations are preferred
c — chemically resistant formulations should be used

Applications of moisture-curable polyurethane sealants in the building industry

138

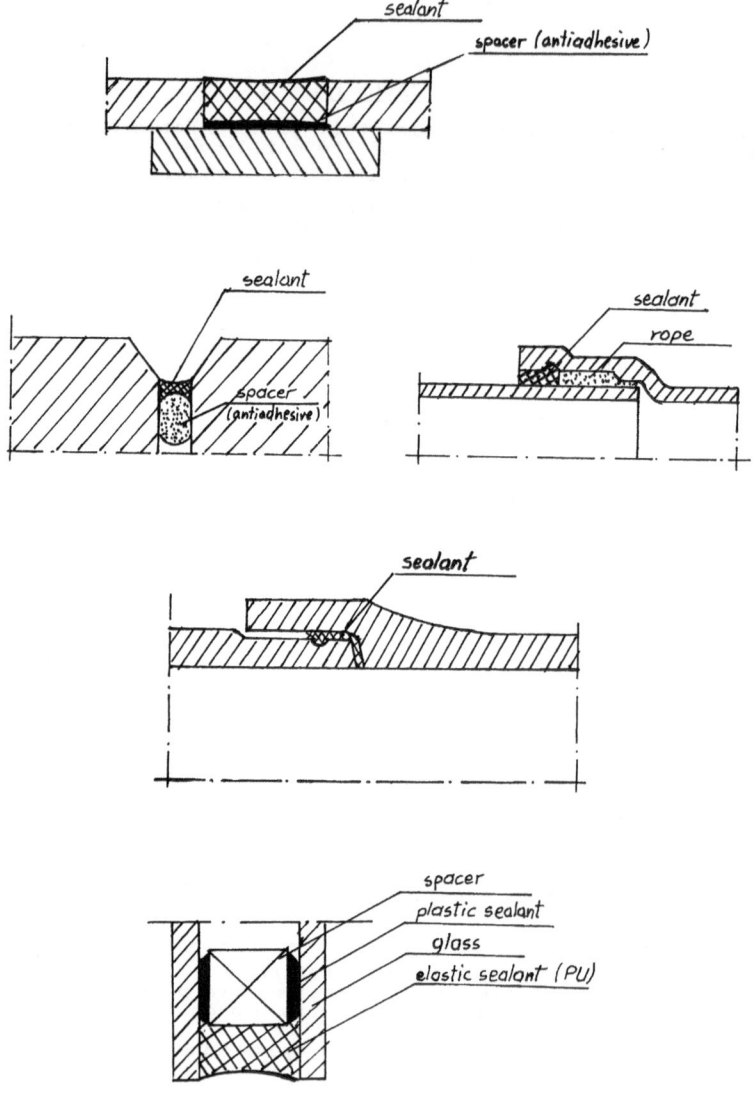

Fig. 7   Examples of joints where moisture curable PU sealants
can be applied.

139

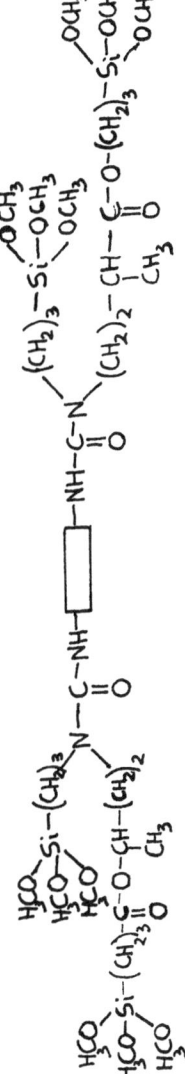

Fig. 8   Structure of silane-terminated urethane prepolymer –
potential raw material for building sealants.

☐ – urethane prepolymer backbone

ideas for producing tailor-made structures, sometimes combining the best properties of chemically different products, the reaction between NCO and functional groups of certain silanes being the most popular way[56-61]. As an example the structure of the branched silane-terminated urethane polymer recommended as the raw material for moisture-curable building sealants[61] is presented in Fig. 8.

References

1. Pletzke T. Adhes. Age 29/(5), 22,1986.
2. Adhes. Age 29/(11), 17,1986.
3. French E.L. in: Alner D.J.ed. "Aspects of Adhesion" 1, 120. University of London Press Limited, London 1963
4. Waters W.T.: Adhes Age 29/(1), 19,1986.
5. Lucke H.: Adhasion 10, 364,1970.
6. Kozakiewicz J., Proc of IVth School of Polyurethanes, Tuzla, 1-6 Sept. 1986 p.185.
7. Shitov V.S. et al. : Kauts. i Resina No. 7,1981.
8. Frommelt S., Rubner J.: Syspur-Reporter No. 13,1976, p.1, lecture presented at II International PUR Symposiums zur Leipziger Herbstmesse 1976.
9. Japan Kokai Tokyo Koho 80-55863; CA 93 133539.
10. Europ. Pat. Appl. 19159; CA 94 122766.
11. Kuksin A. : Proc of VIIth Conference of Metal Bonding "Intermetalbond 81", pp.53 and 62, Bratislava 1981.
12. Wen-Hsuan Chang et al: Ind Eng Chem Prod Res Develop. 12(4),278, 1973.
13. Jap. Pat. 59 131 683: CA 101 231639.
14. Kujawa-Penczek B., Penczek P. : Adhesion 28(3), 7, 1984.
15. Kujawa-Penczek B. Mielniczuk E.: Polimery/Warsaw in print.
16. Kotwica I., Kujawa-Penczek B., Mielniczuk E., Siemieniako I. : Adhasion 31(7-8), 26, 1987.
17. Mielniczuk E., Kujawa-Penczek B. : Poster presented at the Conference of Polish Chemical Society, Bydgoszcz, 3-5 Sept. 1987.
18. Kotwica I., Kujawa-Penczek B. : Adhasion in print.
19. River B.H.: Adhes. Age 27(2), 16, 1984.
20. Voithenberg H. : Prepr. of VIIIthe Munich Seminar on Adhesives and Coatings, p.25, 1983.
21. Knopf B. : Kunststoffe 76, 783, 1986.
22. Evans R.M. : Proc. of IVth School of Polyurethanes, Tuzla 1-6 Sept. 1986, p.103.
23. Gruber H. Farbe u. Lack 80(9), 83, 1974.
24. Brit. Pat. 1 031 917.
25. Brit. Pat. 1 200 718.
26. US Pat. 3 267 078.
27. US Pat. 3 932 257.
28. Brit. Pat. 1 463 408.
29. US Pat. 3 372 371.
30. US Pat. 4 067 844.
31. Ger. Pat. DE 3 133 769; CA 98 199971.
32. US Pat. 3 865 791.

33. Schering Informational Data Sheet, 1984.
34. French Pat. 2 482 964.
35. Ger. Pat. Appl. 2 458 588.
36. Kozakiewicz J., Orzechowski A., Lendzion A., Raszczuk A. : Polimery, Warsaw, 27, 266, 1982.
37. Kozakiewicz J., Orzechowski A., Lendzion A. : Proc. of Rubber 84 Conference, Moscow 1984, Preprint C-89.
38. Jap. Pat. 75 04199.
39. US Pat. 3 711 345.
40. Ger. Pat. 3 129 400 ; CA 98 162619.
41. Pol. Pat. Appl. P-258 346.
42. Bukowski A., Gretkiewicz J. : Appl Polym. Sci 27, 1197, 1982.
43. Bukowski A., Gretkiewicz J. : Plaste u. Kutsch. 28/2, 86,
44. Kozakiewicz J., Lendzion A. : Chapt. 9 in Allen K.W. ed. "Adhesion 9", Elsevier Appl. Sci. Publ., London, New York 1984, p.134.
45. Blomeyer F. : J. Oil Col. Chem. Assoc. 55, 977, 1972.
46. Kozakiewicz J., Wlazlo A. : Full Texts of Int. Rubber Conf., Kyoto 1985, paper No. 17 A 04, p.399.
47. Kozakiewicz J., Wlazlo A. : Kautsch. Gummi Kunststst. 40, 136 1987.
48. Kozakiewicz J. : Proc. of IVth School of Polyurethanes, Tuzla, 1-6 Sept. 1986, p.174.
49. US Pat. 3 372 083.
50. Breech J.C., Turner C.H.C. : J. Chem. Technol. Biotechnol., Chem. Technol. 33A(1), 63, 1983.
51. Orzechowski A., Kozakiewicz J., Lendzion A. : Polimery, Warsaw 29, 69, 1984.
52. Kozakiewicz J., Orzechowski A., Lendzion A., Krowicka I. : Mat. Budowl. No. 1, 19, 1983.
53. Orzechowski A., Kozakiewicz J. : Proc. of Rubber 84 Conference Moscow 1984, Preprint A 105.
54. Pol. Pat. 133 783.
55. Raszczuk A., Kozakiewicz J., Orzechowski A. : Appl. Poly. Sci. 31, 135, 1986.
56. US Pat. 3 632 557.
57. US Pat. 3 711 445.
58. US Pat. 3 979 344.
59. US Pat. 4 067 844.
60. US Pat. 4 222 925.
61. US Pat. 4 374 237.

10

COMPOSITE TO METAL JOINTING FOR TRANSPORT APPLICATIONS

R.J. LEE, R. DAVIDSON & J.C. McCARTHY*
Materials Development and Engineering* Divisions
Harwell Laboratory,
Oxon. OX11 0RA

INTRODUCTION

Advances in the design, manufacture and use of composite materials as
lightweight alternatives to all-metal construction has led to an increased
awareness of the need for efficient structural jointing methods. Interest
in the use of these materials for automotive applications [1] continues
and a recent review describes the current status and future prospects for
composite springs, driveshafts and engine components [2]. The
introduction of advanced composites raises a number of design problems at
joints such as anisotropy, thermal expansion mis-match, electrical
continuity and environmental durability. Structural bonding has been
common practise in the aerospace industry for over 40 years and is widely
used in fuselage and wing construction as a method of attaching stiffeners
to metal skins.

This paper describes part of a research programme to develop improved
design procedures for bonded composite to metal joints. It was sponsored
by 19 companies from European automotive, aerospace and materials supply
sectors, the UK government and the EEC and ran from 1981 to 1984.
Contractual obligations meant that no information was permitted to be
published for at least three years from the end of the programme. The
overall aim of this work was to promote the use of lightweight and
structurally efficient composite materials in transport applications.
This would lead to weight reduction with a consequent decrease in fuel
consumption. Other improvements in structural or environmental parameters
such as noise, corrosion and impact absorption could also result.

The general approach was to provide user friendly computer codes and to
produce relevant design data on adhesives to show quantitative effects of
surface pretreatments and environmental aging. In addition component
testing was performed to focus the fundamental work on how it can be
applied to realistic components.

The testing programme covered the following:-

-       Selection and characterisation of candidate adhesives.

142

- Development of a suitable test methodology to provide reliable data on adhesive properties.

- Evaluate environmental durability and surface pretreatments for the combinations of adherends selected for the joint components.

- Mechanical testing of full size component joints for driveshafts and structural brackets for static and fatigue loading.

- Assessment of a suitable non-destructive examination method compatible with the anticipated volume of structural joints used in mass production.

MATHEMATICAL MODELLING

The analysis of adhesive bonds is complex due to the non-linear function of materials properties and bond geometry. The ability to predict joint strength, however, is central to the design function in order to eliminate costly development exercises and establish confidence in this method of assembly. A survey of existing design techniques for adhesive bonding revealed that a model based on a continuum mechanics approach was most likely to provide a design technique which could be widely applied and used for parametric studies of adherend type, geometries, adhesive properties and thermal effects.

Two computer codes were developed for the programme. One for analysing coaxial, tubular, joints, which ignored peel stresses (which are small for this geometry) and a second code for analysing a two dimensional slice through a lap-joint subjected to any combination of tensile, shear or bending loads. The first program was given the name of BISEPS-TUG (Bonded Inelastic Strength Prediction Suite - Tubular Geometry) and the second BISEPS-LOCO (Bonded Inelastic Strength Prediction Suite - Lap Joint Combined Loading).

BISEPS-TUG is capable of analysing coaxial tubular joints using non-linear adhesive properties to estimate shear stresses and strains within the joint with the intention of estimating the load carrying capacity. It is capable of doing this for torsional or axial loads, although for axial loads small peel components are present which are ignored. The features offered by this micro computer based code are:-

- axial or torsional loads

- thermal strains

- adhesive non-linearity

- geometric profiling of adherends

- variable bondline thickness

BISEPS-LOCO was developed to analyse sheet to sheet bonds with combined tensile, shear or bending loads. The code is believed to be unique because of its capability for performing non-linear analysis of both tensile and shear stresses within the adhesive under fully interacting conditions and could therefore be applied to a very wide range of problems. Its main attributes are:-

- sheet to sheet bonds

- tensile, shear or bending loads

- thermal strains

- anisotropic adherends

- peel and shear stress/strain prediction

- non-linear analysis.

The programme was written in Fortran and although developed originally for use on a mainframe also runs on a personal microcomputer.

MATERIALS TESTING

In order to design structural joints with confidence it is essential to know details of the mechanical and chemical characteristics of adhesives. Factors which are of major importance include:-

- stiffness and strength of adhesives

- surface pretreatments of adherends

- resistance of adhesive bond in warm, wet environments

- adhesive application and drying conditions, which are especially important where large volume production is required.

Investigation of these factors was done using simple coupon testing, made from a variety of metallic and composite adherends, and subjected to different pretreatments and environmental ageing.

Adhesives were selected, as typical examples, from commercially available products from each particular class of structural adhesive. In some cases two different adhesives from the same class were chosen to reflect the different loading situations encountered in the two generic types of joint examined, for example, where significant peel stresses are present then adequate 'toughness' may be important, while for a joint subjected to shear loads then creep resistance may be a greater requirement. Table 1 lists the adhesives investigated.

In the following sections various test methods employed are described, in brief they are:-

- Thick adherend shear test (TAST) to characterise the mechanical properties of the adhesives in shear. Some comparative butt torsion results were also included.

- Boeing wedge testing was selected for accelerated environmental durability studies and to indicate the likely performance of deliberately contaminated surfaces.

TABLE 1
Adhesives evaluated

Adhesive	Generic Type	Manufacturer
ESP 105	One part toughened epoxide	Permabond
2007	One part toughened epoxide	Ciba-Geigy
2005	Two part epoxide	Ciba-Geigy
AY105/HY932	Two part epoxide	Ciba-Geigy
F241	Toughened acrylic	Permabond
XB3088	Toughened acrylic	Ciba-Geigy
Pliogrip 6040/6046	Two part polyurethane	Goodyear
Redux 80	Nitrile phenolic film	Ciba-Geigy
Redux 108/308	Epoxide film	Ciba-Geigy

-   Single and double lap shear testing to investigate the
    durability of combinations of composite and metallic adherends.

-   Bulk samples of cast adhesive were subjected to dynamic
    mechanical analysis (DMA) to provide comparative storage and
    loss moduli over a wide temperature range.

THICK ADHEREND SHEAR TEST

The design of this specimen is such that if the adherends are rigid, and
the overlap length is short then loads required to fail the adhesive do
not deform the adherends and the shear strain distribution along the
bondline is reasonably uniform. Its main criticisms are that it is not a
pure shear test as there are small peel stresses at the joint ends and the
square cut in the adhesive at each end of the test sections gives rise to
localised strain concentrations which are inherently more severe than
those which occur in a correctly designed joint with a well formed natural
spew fillet. An example of a specimen used is shown in Figure 1. In all
cases the adherends used were high yield strength steel which were
prepared for bonding by dry grit blasting and a solvent degrease and in
some cases a silane primer was applied. The adherends were bonded with
the appropriate test adhesive using four spacer wires per specimen which
were positioned across the specimen width and cured under pressure at the
appropriate cure temperature. The edges of the specimens were polished to
reveal the bondline and slots were cut through the adherends and adhesive
layer to give a 12.5mm overlap length. Adhesive thickness was measured
with a travelling microscope along the central region to be tested. This
method, however, was sometimes inaccurate due to adherend bevelling at the
corners and it was necessary to make measurements of the bondline
thickness after the specimen had been broken by sectioning and polishing
the adjacent edge next to the central region across the specimen width.
Bondline thicknesses were then made optically and averaged across the
specimen width. Specimens were tested in an Instron 1195 universal
testing machine using a pin-loading arrangement at a crosshead
displacement of $5-10 \times 10^{-3}$ m min^{-1}.

After initial, unsuccessful, attempts to measure shear strains using
conventional extensometry an improved, zero gauge length, extensometer was
adapted from a commercially available Zwick DSST extensometer. This had a

Fig. 1   Thick adherend shear test specimen with extensometer mounted.

resolution of ∿ 0.5µm and since it was modified to work with an effective
gauge lengh of zero the adherend deformations were very much reduced. The
extensometer was attached to the specimen such that the measuring pins,
typically 0.75-1.00mm apart, were positioned across the bondline at the
centre of the overlap. The experimental procedure involved measuring the
elastic deformation across the adhesive and then shifting the extensometer
to measure the equivalent deformations of each adherend on either side of
the bondline. Corrections were then applied to the adhesive shear
stress – shear strain curves to take into account adherend deflections,
caused by a combination of shear and elastic rotation. The extensometer
was then repositioned across the bondline and the specimen tested to
failure.

Steel adherends are to be preferred over aluminium since the
correction factors are very much reduced. For a stiff (shear modulus >
1 GPa) epoxide adhesive it is important to take into account the adherend
deformations as these can account for >12% of the measured deflections.
The problems of measuring shear deflections are reduced somewhat for low
modulus adhesives and for thick (> 0.3mm) bondlines. Thin, stiff
adhesive bonds present the greatest problems.

Results obtained from a thick adherend geometry include:-

–   shear modulus

–   average shear strength

–   average shear strain to failure

      —   an estimate of the onset of non-linearity.

Figure 2 shows results obtained from a toughened epoxide which indicate that, once corrections are applied, shear modulus appears approximately constant at 1.10 $\pm$ 0.1 GPa with bondline thickness ranging from 0.05 to 0.55mm. Figure 3, however, shows that observed strain to failure decreased steadily with bondline thickness over this range.

BUTT TORSION TEST

As an independent verification of the thick adhered shear test some ambient temperature measurements were carried out on the same batch of adhesives using torsional facilities at RARDE (Christchurch) [3]. Torque-twist curves were obtained using HE30 aluminium alloy adherends used in conjunction with a high resolution twistometer. Typical curves obtained are shown in Figure 4 together with the response of an unbonded calibration bar. In general the shear modulus values were in good agreement with those obtained from thick adherend shear testing.

DYNAMIC MECHANICAL ANALYSIS

Obtaining accurate, reproducible, values for shear modulus in very stiff adhesives (G >1.5 GPa) was particularly difficult using the standard TAST geometry. Dynamic mechanical analysis was used to provide another means of obtaining quantitative stiffness data on adhesive films over a wide temperature range. Rectangular section samples of cured adhesive were mounted between the arms of a Du Pont model 1090/982 thermal analyser/DMA and electrically oscillated at the sample's resonant frequency. The data were used to calculate storage and loss moduli over a wide temperature range of −160°C to +150°C and stiffness results were compared with corresponding thick adherend shear results, Figure 5.

BOEING WEDGE TEST

The Boeing wedge test [4,5] was investigated because it offered the possibility of an accelerated environmental durability test. The test method originated at the Boeing Commercial Airplane Company where traditional environmental testing, such as lap shear and peel tests exposed to warm humid environments without pre-stress did not satisfactorily replicate the failure characteristics observed in aircraft components in service. The specimen is designed to measure the adhesive's resistance to cleavage and consists of two bonded adherends with a wedge carefully inserted at one end to maintain a crack opening stress. The stressed specimens were then exposed to an environment of 55°C/95% RH and crack growth data monitored. Emphasis was placed on using metal adherends as environmental degradation was expected to be more severe than with composite adherends. Aluminium alloy and steel adherends were surface treated as shown in Table 2.

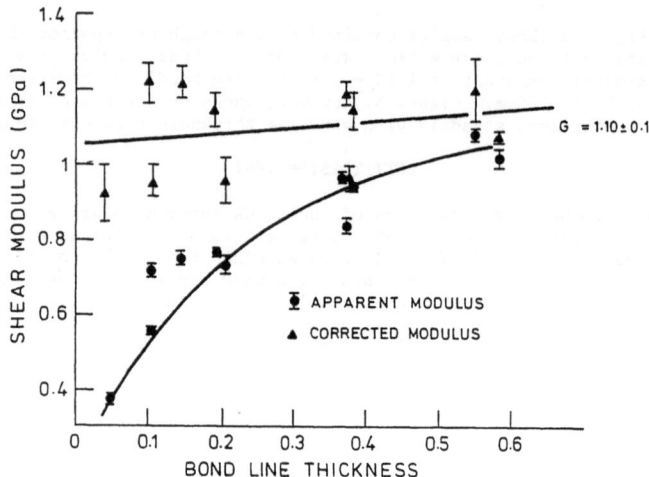

Fig. 2   Variation of shear modulus with bondline thickness for a
toughened epoxide using steel adherends.

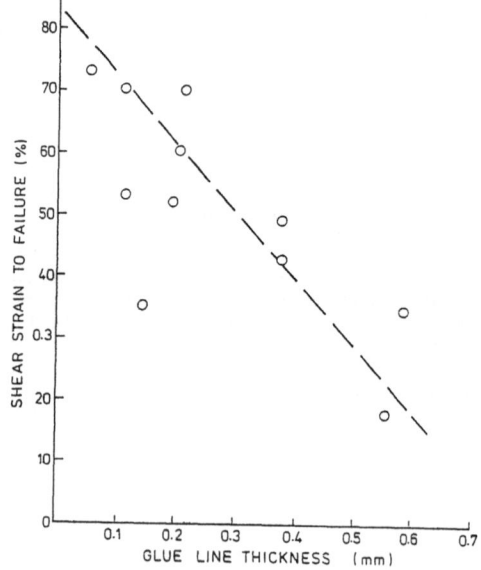

Fig. 3   Variation of failure strain with bondline thickness for a
toughened epoxide using steel adherends.

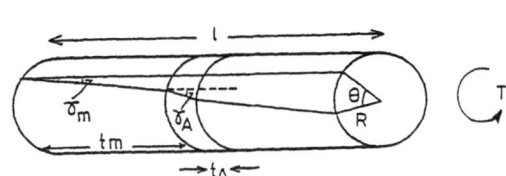

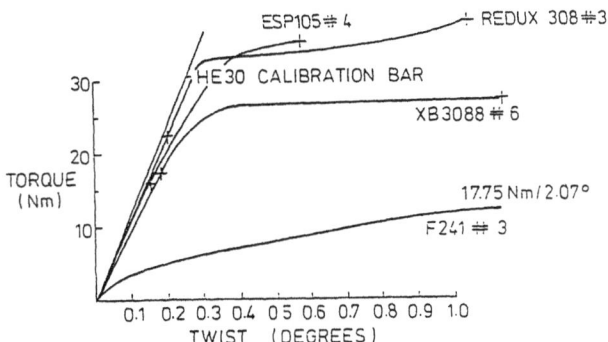

Fig. 4   Typical torque-twist curves for butt torsion specimens.

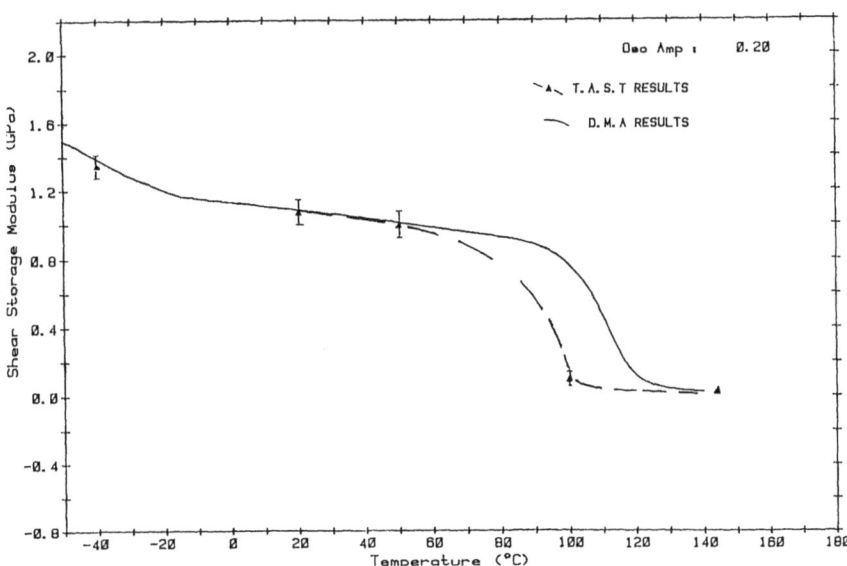

Fig. 5   Comparison of thick adherend shear and DMA results for a
toughened epoxide.

TABLE 2
Materials used for wedge testing

Adherends	Adhesives	Surface Treatments	Environment
Aluminium (LM25)	2007 AY105/HY932		
Steel (EN3B '20' Carbon)	XB3088 Redux 80S	Etched Abraded Contaminated	95% RH 55°C
Aluminium (HE30)	ESP105 2005	(Silane for Steel)	
Steel (EN3B '20' Carbon)	F241 Redux 108/308		

The results can be presented as crack growth against time, or can be analysed to give fracture energy as a function of time. In this case by equating the stored elastic energy in the cantilever beam to the fracture energy:

$$F = \frac{Ed^2h^3}{16} \frac{[3(a+0.6h)^2+h^2]}{[(a+0.6h)^3+ah^2]^2}$$

where E is the adherend modulus
     h is the adherend thickness
     t is the adhesive thickness
     a is the crack length from bearing edges of wedge to crack tip
     s is wedge thickness
     d is the displacement of the adherend by the wedges (d=s-t).

Figure 6 shows a typical example of data obtained with on steel adherends.

It is worth commenting that, although this is a good screening test for adherend surface quality, it has a number of problems such as reproducibility of the initial crack lengths after wedge insertion. Since the stress concentration at the crack tip depends upon the length of crack, the driving force for crack growth will not be constant from one specimen to the next and will also vary for a given specimen as the crack extends. An added complication, particularly for steel adherends, is corrosion of the adherend and the added difficulty of observing the crack tip. The test is only semi-quantitative because of plastic deformation of the adherends, particularly when aluminium alloy adherends are used in conjunction with tough adhesives.

LAP SHEAR TESTING

The lap shear test is not recommended for producing quantitative design data but is useful for comparing the relative durability of selected joints. Single and double lap joints were tested using a mixture of metallic and composite adherends. The aim of these tests were to compare joint durability with different pretreatments and adhesives.

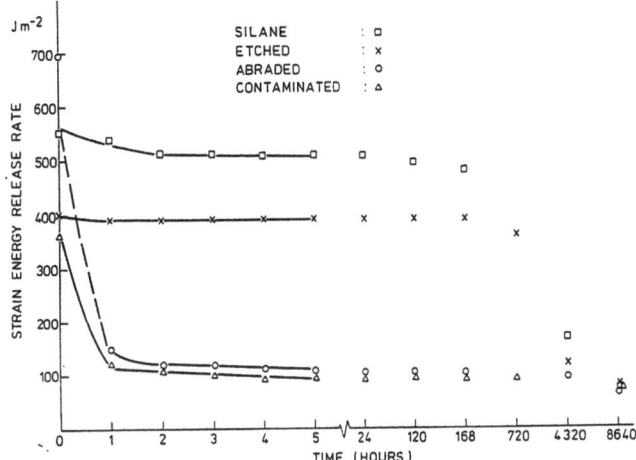

Fig. 6   Boeing wedge data for a toughened epoxide on steel adherends.

Surface pretreatments chosen were:

1. **Etched**
   An aerospace quality finish. For aluminium this followed the
   Phosphoric Acid Anodise (PAA) treatment in Boeing specifications
   BAC 5514 & 5555; for steel the phosphoric acid etch described in
   BS 5350 was used.

2. **Abraded**
   Surfaces were dry grit blasted and degreased.

3. **Contaminated**
   Light mineral oil was smeared over the surface after grit
   blasting and degreasing.

4. **Silane primer**
   For steel adherends an aqueous solution of a silane adhesion
   promotor was applied
   (Union Carbide A187 : γ glycidoxypropyltrimethoxy silane).

   Traces of release agents used in the fabrication of composite
laminates were detected as fluorine or silicon on the composite surface
using X-ray photoelectron spectroscopy, even after the removal of peel
plies. Mechanical abrasion followed by a Genklene wipe proved an
effective means of removal of the surface contamination. Sheet moulding
compounds (SMC) use zinc stearate as an internal release agent and this is
difficult to remove. In the case of SMC, however, this is not such a
serious problem as the interlaminar shear strength of SMC is often the
limiting factor in joint strength.

An example of results obtained for double lap shear specimens of CFRP/GRP hybrid bonded to steel with an epoxy adhesive and using a 5 $^w/_o$ solution of A187 silane applied to freshly sand blasted surfaces are shown in Table 3.

TABLE 3
Ageing of composite/steel double lap shear specimens

Ageing	Double lap shear strength (MPa)	
	Untreated	$5^w/_o$ A187 silane
None	28.6 ± 5.3	22.7 ± 4.39
3 months 21°C/100% RH	17.6 ± 5.4	31.5 ± 1.45
3 months 60°C/100% RH	5.8 ± 0.35	10.8 ± 3.75

COMPONENT TESTING

Component testing of model joints was undertaken to focus the attention of the programme on problems of genuine concern. Specifications were provided by sponsors from the automotive industry and took into account the fatigue loading spectrum and operating environment. Joint geometries and materials were chosen to reflect specific applications for composite to metal joints. These had particular emphasis for road vehicles but may also be relevent to aerospace structures. The components selected were:-

(1) Components subjected to torsion loading

   (a) A front wheel drive car driveshaft, capable of transmitting a maximum torque of 1000 Nm, employing a filament wound carbon fibre tube with steel end fittings.

   (b) A truck propeller shaft with a torque requirement of 5000 Nm using a filament wound hybrid tube consisting of glass and carbon fibre with aluminium alloy end fittings.

(2) Composite bracket components

   (a) A chassis bracket for mounting fixtures on the structural members of a truck using a glass/polyester sheet moulding compound (SMC).

   (b) A hatch-back door bracket for hinge attachment, also in SMC.

CARBON FIBRE DRIVESHAFT WITH STEEL END FITTINGS

Specifications were provided by Working Party members for a wide range of driveshafts, particularly for front-wheel-drive car applications. A study of these revealed that maximum load conditions placed fairly high stresses on the the existing metal components although these load conditions were infrequent and the dynamic nature of the loading and the freedom of the shaft to accelerate would tend to cause a reduction in these load conditions. Steel components often have a toughened outer layer or

153

prestressed compression on the outer surface, which tends to inhibit the propagation of fatigue cracks. The substitution of composite materials in these components would need to be similarly optimised and take account of the true loading spectrum including thermal and secondary stresses.

Carbon fibre tubes were filament wound at Harwell using continuous carbon fibre tow laid onto a mandrel at ± 45° and made in 1.55 m lengths which were subsequently cut to test lengths of 300 mm. Early tubes were shell wound, later tubes were of basket weave construction. Off cuts from preparation of the test lengths were polished and examined by optical microscopy as a measure of tube quality.

Adhesives were chosen on consideration of the results of coupon tests and quality control was checked by lap shear tests. Steel end pieces were made in two configurations. In both designs torque was applied through 50 mm long by 34 mm square section ends with radiused corners and cylindrical bonding sections 65 mm long, inserted into the carbon fibre tube. A reference design had a bonding section that was a simple solid cylinder, the other, profiled, had material removed from the core of the bonding section to taper the thickness of steel. The design of the profiled joint was made using the BISEPS-TUG computer program. The result of profiling is to minimise stress concentrations in comparison with a reference joint and the shear strain distribution of the two designs as computed by BISEPS-TUG are shown in Figure 7.

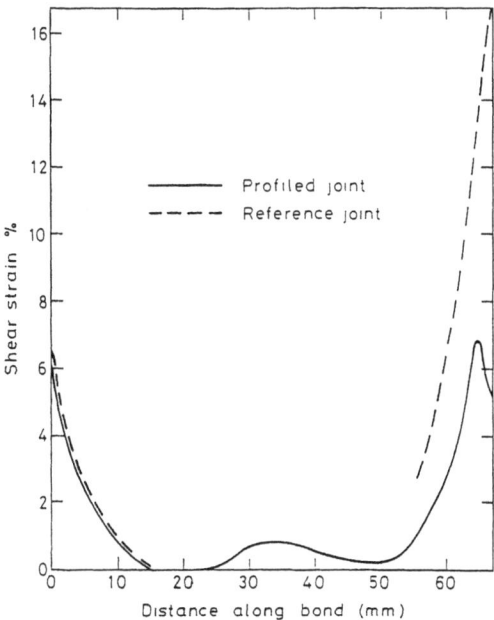

Fig. 7   Distribution of shear strains in CFRP driveshaft.

Test results comparing profiled with reference joints are given in Table 4.

TABLE 4
Static strengths of CFRP driveshaft using an epoxide adhesive

Joint Design	Test Temperature (°C)		
	60	20	150
Profiled	900 ± 61	926 ± 187	630 ± 20
Reference		757 ± 222	

A number of specimens that failed at the joint were sectioned and examined. The bulk of failures appeared to be due to interlaminar shear near the inner surface of the tube where the load is transferred from the end fitting via the adhesive to the composite tube.

Fatigue test results (carried out under load control) are shown in Figure 8. Three further tests were carried out at 0-400 Nm loading on specimens with profiled ends that had been subjected to 10 cycles of thermal cycling between 100°C and -60°C; one survived 10^6 cycles of 0 to 400 Nm loading and had a residual strength of 972 Nm while the second failed at 810,000 cycles and the third after 13050 cycles with a final static strength of 525 Nm.

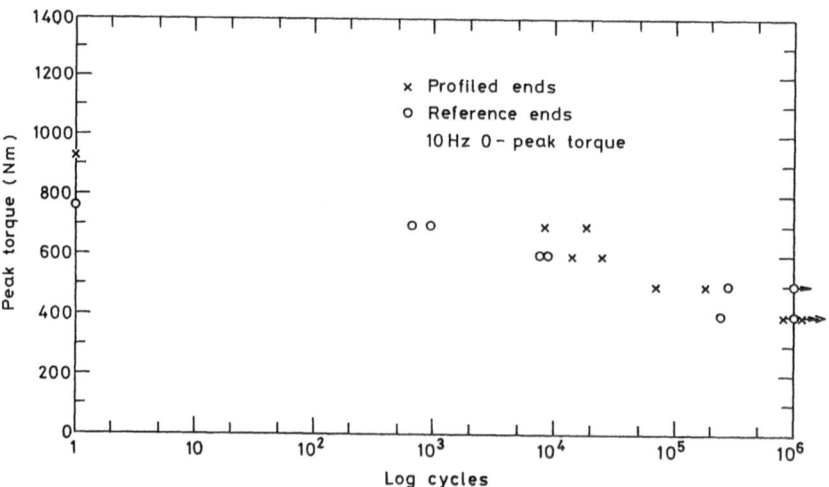

Fig. 8   Fatigue results for CFRP driveshaft.

Carbon/Glass Fibre Hybrid Propeller Shaft with Aluminium End Fittings

The design of a composite shaft which met the requirements with the lowest material cost, whilst still giving substantial weight savings, used a glass fibre composite wound at +/- 45 degrees. The addition of axial carbon fibre was necessary to allow the specification for critical whirling speed to be met. A volume ratio of glass to carbon of 7:1 was selected although for longer shafts more carbon is needed.

The resulting composite hybrid design gave a weight saving of about 30% whilst meeting the required specification. This weight saving could be further improved if a larger diameter were allowed in the specification.

Aluminium castings (LM25) were machined to produce reference and profiled end fittings. Some Duralumin profiled end fittings were also obtained for comparison purposes. Figure 9 depicts a BISEPS-TUG comparison of projected shear strains in the adhesive layer for the reference and profiled fittings.

Fatigue test results are given in Figure 10 which indicate that the profiled joints showed lifetimes in excess of 10^6 cycles at loads above the static specification while the reference joints had lifetimes of 10^6 cycles above the 40% fatigue specification.

COMPOSITE BRACKET COMPONENTS

The general types of composite-metal joint under consideration in this part of the project were for brackets where the joint is subjected to bending and shear loading. The various joining options covered both mechanical fixing and adhesive bonding including combinations of these two methods. For joints subjected to relatively low stresses the unit cost of the joining method would be a dominant design consideration but for higher stresses the structural integrity is also of prime importance.

The poor fatigue performance of mechanical fixings and the cost of required moulding inserts makes this method less attractive for composites than for metal-metal joints. On the other hand the extreme modulus mismatch between, for example, a glass composite and steel makes the load distribution through an adhesive bond very uneven unless due consideration is given to designing the joint carefully. In addition the poor performance of adhesives against peeling stresses makes the use of combined mechanical and adhesive bonding desirable in some cases. For these reasons the bracket/top hat type of joint formed a challenging exercise in design and a useful background for the comparison of joining methods. Two types of joint were considered in detail, a chassis bracket and a hatchback door attachment bracket.

LORRY CHASSIS BRACKET

A typical specification for a bracket found on the longitudinal structural members of a typical truck was selected. Such brackets are used to mount most of the fixtures on the vehicle including the cab, engine, suspension fuel tank and battery.

This type of bracket is subjected to considerable peel loading due to acceleration forces in cornering and it was thought that mechanical

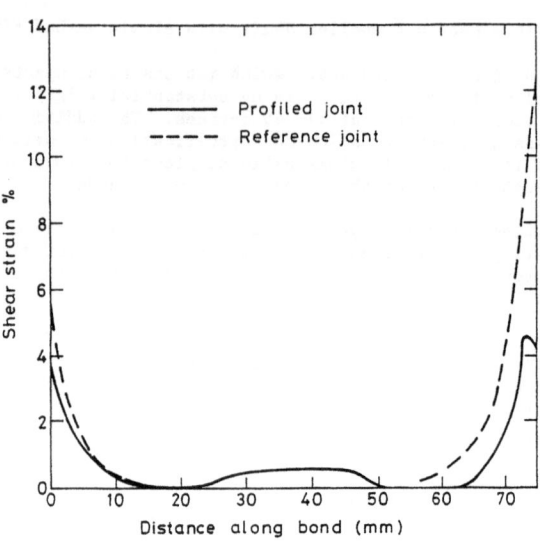

Fig. 9   Distribution of shear strains in propeller shaft joints.

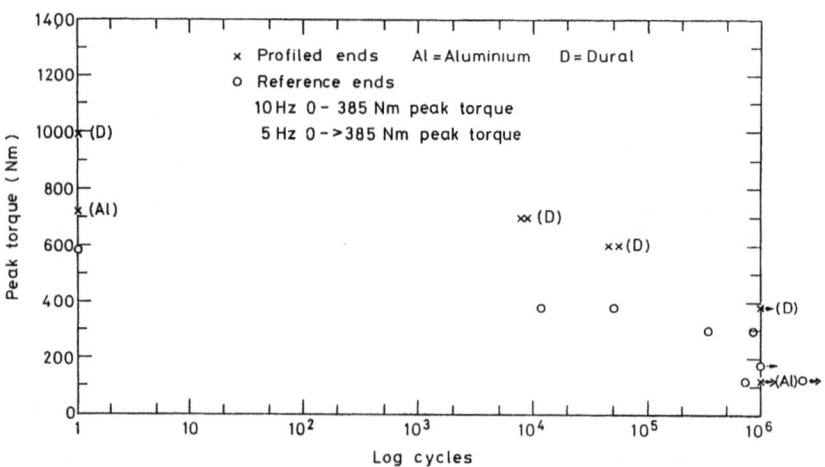

Fig. 10   Fatigue results of scaled down hybrid propeller shafts.

fixing in the form of bolting or riveting would be an additional requirement. Pure mechanical fastening is often inadequate due to poor fatigue performance and tendency to provide stress concentrations. The combination of the two joining modes has proved in the aerospace industry to perform very much better than either of the modes used separately.

Typical loading magnitudes and directions are given in Figure 11. These loads are the maximum values encountered as a result of a suspended mass at the illustrated loading point being accelerated by bump loading for the vertical load case and cornering loading for the horizontal load case.

Typical spatial limitations required that the width of the joint was 120 mm. The maximum depth of the section was also taken as 120 mm to prevent excessive protrusion from the side of the vehicle. The metal structural channel section had to remain clear at the top and bottom flanges and mechanical fastening was only allowed if holes made in the metal component were near the centre line or neutral axis to minimise any strength reduction of the structural member.

The bracket eventually selected was manufactured from 40 weight percent glass sheet moulding compound and is shown in Figure 12.

The maximum wall thickness of the glass fibre composite was 6 mm for ease of moulding and was profiled in depth over the area of the joint. Full use of the available space envelope was made because of the high bending loads. The temperature range of the extremes of the local environment in the region of the bracket were taken to be from −40 to +120°C.

Preparation of the steel channel was by sand blasting followed by degreasing prior to application of the acrylic adhesive. The SMC surface was roughened using an abrasive pad, degreased and the initiator applied. Bolts were treated with a mould release compound and tightened to a specified torque. Specimens were left at least 24 hours at room temperature before testing took place.

Testing was carried out either parallel to the SMC channel − load case I or orthogonally to the SMC channel − load case II.

The mode of failure induced in loading case I on earlier channel designs led to collapse of the SMC channel and cracking along the bolt lines. Case II produced a peel failure by delamination of the SMC coupled with bolt pull-through. The re-designed SMC channel failed at the open edges of the SMC near the stiffening web for load case I and by bolt pull-through for load case II.

Two load−deflection plots for 40 $^W/_o$ SMC redesigned channels, bolted and bonded with F241 are shown in Figure 13. In both cases there is a linear loading curve to the point where peel fracture of the adhesive occurred. This was followed by a load drop which was followed by a more compliant loading as the bolts were pulled through the SMC.

STRUCTURAL HINGE

A weight saving application for composites is in door frames where the use of SMC is being considered. An added incentive for weight saving is the

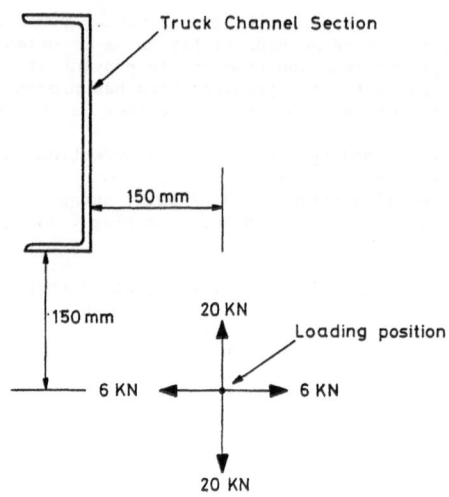

Fig. 11   Specified loads on bracket component.

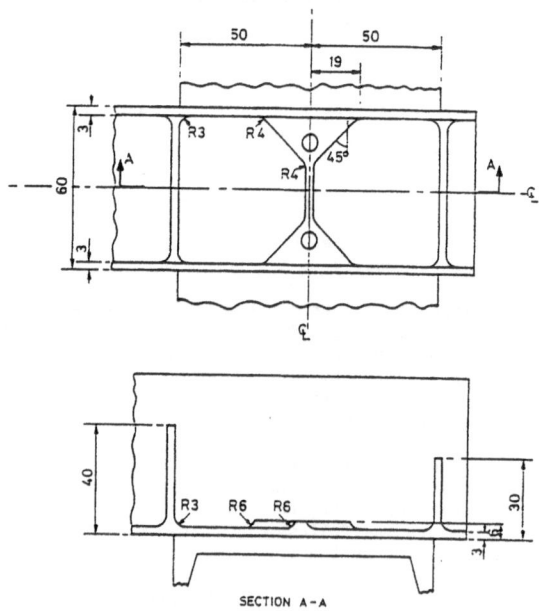

SECTION A-A

Fig. 12   Modified SMC channel/bracket design.

reduction in lifting force required to open a hatchback door. The hinge
attachments for such a door can be difficult to mould and the SMC would
not perform well as a bearing surface. One option for avoiding this
problem is to use a metal hinge attached to the composite by some means.
Mechanical fastening would usually require moulded inserts in the SMC to
avoid any unsightly feature on the outside of the door. However, moulded
inserts would lead to additional cost which might be avoided by the use of
adhesive bonding.

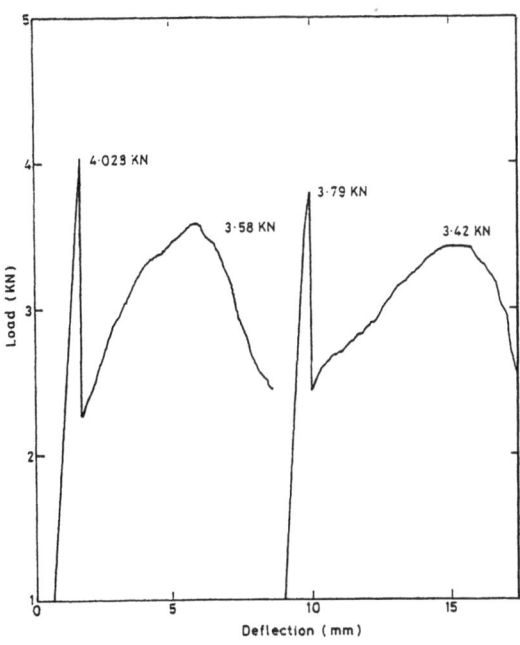

Fig. 13   Results of modified bracket assembly bolted and bonded loaded in
           Mode II.

The hinge is subjected to forces during assembly, opening and closing
operations and road usage. The latter forces arise from bump conditions
where the body shell distorts and the force is transmitted through the
hinges and the rubber seal to the body shell as the door resists
deformation in its own plane. The door is subjected to a large number of
opening and closing operations and the test specification requires that
the door is capable of 20,000 cycles at a closing speed of 1.6 m/s at the
level of the lock.

The design for the hinge was envisaged as a slot into which a
pre-heated steel hinge could be pushed. Adhesive would be placed in the
slot prior to insertion of the hinge thus ensuring a rapid cure without
heating the SMC to high temperatures and minimising moisture boiling from
the SMC causing voidage. A further alternative was the use of a compliant
room temperature curing adhesive.

Although, in final use, hinge slots would be moulded into an entire SMC door frame, for test purposes an attachment design was adopted which had the additional possibility of being bolted to a modified metal hatchback door for slam test trials.

The SMC specification chosen was 25 $^w/_o$ glass fibre weighting in a filled polyester resin with a cure temperature of 140°C. A high quality 1.5% Mn, 0.2% C weldable structural steel was used for the insert material.

Some adhesive specimens were quench cured: the adhesive was inserted into the SMC slot and hinge steel, pre-heated between 200 and 250°C was inserted. This was usually followed by a post cure of 1 h at 150°C. For other adhesives, curing occurred at 80°C for 1 h and the effect of a post cure at 150°C for 1 h was investigated. Similarly, a room temperature cured adhesive was tested with and without the same post curing.

Tests were carried out initially in two modes, both having offset loading to simulate actual geometries in a door. Mode I parallel to the metal insertion direction while mode II was orthogonal. During testing, it was found that most of the deflection was due to bending of the two sheet metal pieces. The curved piece of sheet steel representing the hinge was replaced by a solid block that maintained the required testing geometry. Considerable bending of the steel inserted into the SMC took place before damage was sustained by the SMC.

Static test results in modes I and II are given in Table 5 include tests at 100°C and -30°C.

Two SMC brackets were bonded to metal hinges and attached to a modified hatchback door for subsequent slam testing on a body shell. Tests of the door's rigidity and endurance on a slam rig were performed by one of the working party members.

The assembly successfully withstood 20,000 cycles in the slam rig. Its rigidity was greater than the conventional door with welded steel hinges, but this is partially due to the changes made to the top of the door to accommodate the SMC bracket.

SUMMARY

Two computer programs were developed for analysing adhesive joints.

BISEPS-TUG predicts the stress in a coaxial tubular joint subject to torsion or axial loads and thermal stress. It can take account of plasticity in the adhesive.

BISEPS-LOCO predicts the stress in a lap shear joint subject to any combination of tensile, shear or bending load and thermal stress. It can take account of plasticity in the adhesive.

Coupon testing, using thick adherend shear tests (TAST), wedge tests and lap shear tests of candidate adhesives, proved to be a valuable means of determining the shear strength and stiffness characteristics, and durability of structural joints.

TABLE 5
Static tests on hinge brackets

Joint Details	Load (KN)	Failure Mode
Specification Mode I	0.25	
Specification Mode II	2.0	
Quench + post cure Mode I	2.95	SMC breaking, steel bending.
Quench + post cure Mode II	4.8	SMC breaking, steel bending.
Quench + post cure Mode III	5.0	Steel bending.
Mode I	1.1	SMC cracked.
Mode II	4.1	Steel bending.
Quench + post cure Mode I (100°C)	0.625	SMC failed.
Quench + post cure Mode I (-30°C)	1.82	SMC cracking.
Quench + post cure Mode II (96°C)	3.92	SMC cracking and steel bending.
Quench + post cure Mode II (-49°C)	4.8	Steel bending.

Dynamic mechanical analysis testing on bulk samples of adhesive has been shown to be a powerful and rapid technique to determine shear moduli and damping characteristics over a wide temperature range. DMA also has been shown to give a means to investigate the effect of cure cycle on the thermal characteristics of adhesives.

Single and double lap shear testing on composite/metal joints gave results in broad agreement with the metal/metal wedge test results. An A187 silane primer was the most effective treatment on abraded steel adherends. The lap shear data were distorted to some extent by the fact that failure often occurred within or very close to the composite adherend. This tends to disguise the influence of the metallic surface pretreatment.

The CFRP torsion drive shaft joints met the design specification.

The hybrid glass–carbon propeller shaft exceeded the static specification by a factor of 5 with the aluminium adherends being the weak point.

The SMC bracket channel with moulded bolt holes and load spreading webs eliminated the use of metal stiffeners and allowed the specification to be exceeded by up to 50% in case I and 100% in case II. The use of adhesive gave a stiffer structure and ultimate failure occurred in the SMC.

The hatchback door hinge bracket specification was easily exceeded and fatigue presented few problems.

ACKNOWLEDGEMENTS

The authors wish to thank members of Harwell's Materials Engineering Centre who participated during the programme. Financial support from the EEC and the Department of Trade and Industry is gratefully acknowledged.

REFERENCES

1. McGeehin, P., Composites in Transportation – Design and Current Developments. Materials in Engineering 3 (April 1982), 378–387.

2. Kretschmer, J., Composites in Automotive Applications– State of the art and prospects. Materials Science and Technology 4 (Sept. 1988), 757–767.

3. Stringer, L.G., Comparison of the shear stress–strain behaviour of some structural adhesives. J. Adhesion 18 (1985), 185–196.

4. Stringer, L.G., Wedge test evaluation of adhesive bonded aluminium alloy joints. MVEE Report 81506 (June 1981).

5. Stone, M.H. and Peet, T., Evaluation of the wedge cleavage test for assessment of durability of adhesive bonded joints. RAE Tech. Memo. Mat. 349 (1980).

11

BONDED JOINT DESIGN ANALYSES

D A BIGWOOD and A D CROCOMBE[*]
Department of Mechanical Engineering
University of Surrey
GUILDFORD, Surrey, GU2 5XH

INTRODUCTION

Recent advances in the formulation of adhesives have enhanced the use of bonding as a structural fastening technique. However, the full potential in the use of modern adhesives has been restricted by the difficulties encountered in evaluating the strength of a proposed joint. These difficulties arise because there is no technique that is readily available to the design engineer for analysing the wide range of possible joint configurations.

In this paper the general plane strain problem of adhesively bonded structures will be considered. A basic approach for the design of bonded joints is outlined including simple formulae which can be used to evaluate the peak adhesive stresses that often occur at the ends of the adhesive overlap. These much simplified formulae will be of considerable use to the engineer in designing joints of simple configurations.

Adhesive joint analysis started as far back as 1938 with the shear-lag analysis of Volkersen [1], which considered the problem of a rivetted joint simplified to be represented by a joint with an elastic interlayer. Volkersen calculated the interlayer shear stress distribution in terms of the differential stretching of the adherends. The analysis is presented as a single-lap joint analysis, but because it ignores the adherend bending effects which predominate in the single-lap joint it is no longer used to analyse this joint but is used to analyse double-lap joint configurations, where bending effects are largely suppressed.

Goland and Reissner [2], produced a single-lap joint analysis which considered the joint as two cylindrically bent plates, joined by an elastic interlayer, and analysed them using plate

---

[*] To whom all correspondence should be sent.

bending theory, modelling both bending and stretching of the adherends.

Since the problem of joint analysis was effectively introduced by these authors various refinements have been made to improve the accuracy of the model. These refinements include: adjusting the single-lap joint boundary conditions and accommodating a limited amount of adhesive plasticity in shear, Hart-Smith [3]; producing a general stepped-lap joint analysis and accommodating non-linear adhesive material characteristics in Volkersen's shear-lag analysis, Grant [4].

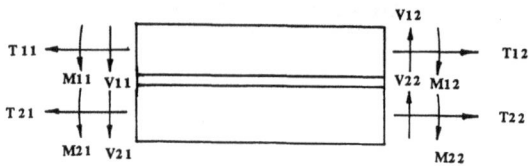

Figure 1a. Simple adherend-adhesive sandwich

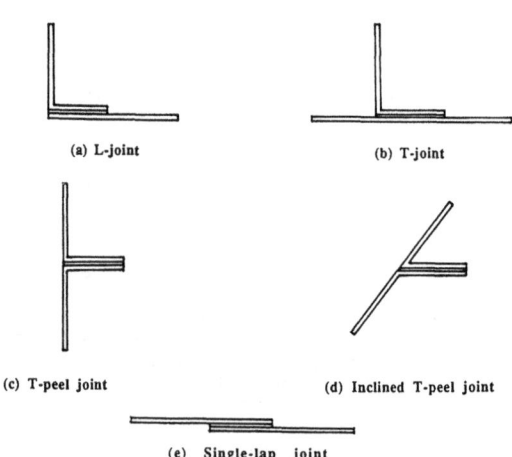

(a) L-joint   (b) T-joint

(c) T-peel joint   (d) Inclined T-peel joint

(e) Single-lap joint

Figure 1b. Joint configurations

Similar basic design approaches have been applied to other joint configurations such as the peel test [5], the scarf and stepped-lap joints [6], and the double-lap joint [7], but a common limitation of all these and similar analyses is that they are 'joint specific' in that they apply to one joint configuration only, further they are not capable of modelling full adhesive and adherend non-linear material behaviour.

In an attempt to rationalise the analysis of bonded joints the overlap region has been reduced to a simple adherend-adhesive sandwich and subjected to general loading as shown in Figure 1a. In such a way joints which can be simplified in this manner, and for which end loading conditions can be found, including those shown in Figure 1b can be analysed.

ELASTIC SIMPLIFIED ANALYSES

Referring to the general elemental diagram shown in Figure 2, two simplified analyses considering either one or other of the two adhesive stress components have been derived. Considering the transverse stress component of the adhesive stress separately, an analysis referred to as the Simplified Peel Analysis has been produced, and considering the adhesive shear stress component separately an analysis referred to as the Simplified Shear Analysis has been produced.

Elastic Simplified Peel Analysis

The adhesive joint is analysed by considering the elemental length of bonded joint shown in Figure 2, which describes the case of dissimilar adherends under general tensile, shear and moment loading per unit width, but the adhesive shear stress, τ_{xy}, is assumed not contribute significantly to the flexural behaviour of the adherends and their associated out-of-plane deformation and is thus neglected. The adherends are analysed using plate bending theory from which the adherend bending is related to the vertical displacement. Force and moment balance equations are derived for each adherend relating the applied loading to the adhesive stress, and lastly a linear transverse stress-strain relationship is assumed linking the stress in the adhesive to the displacements in both adherends by means of a spring analogy.

The above relations are reduced to a single fourth-order differential equation in transverse adhesive stress, shown here:

$$\frac{d^4\sigma_y}{dx^4} + K_3'\sigma_y = 0$$

(1)

with a solution of the general form:

$$\sigma_y = A_1 CoshmxCosmx + B_1 CoshmxSinmx + C_1 SinhmxCosmx + D_1 SinhmxSinmx$$

(2)

Using four boundary conditions relating combinations of loads at the ends of the sandwich to solve for constants $A_1 - D_1$ in equation (2) explicit expressions for the transverse y-direction stress, σ_y, can be produced in a similar fashion to analyses like Goland and Reissner [2] but for a general sandwich configuration.

Simplifications can be made to the analysis to produce the expressions shown in Figure 3, describing the peak transverse stress in the adhesive layer occurring at the left end of the sandwich. Firstly, the overlap is assumed to be long, for a typical epoxy-aluminium joint with thicknesses of 0.25 and 2.0mm respectively, this implies a minimum overlap of about 10mm. Secondly, each load component at the left end of the overlap (i.e. at x = 0) is considered separately. In this way peak stress equations can be expressed very simply in terms of two variables, called the peel compliance factors, which are a measure of relative adherend-adhesive stiffness, and are defined in Figure 3. Any combination of shear and moment loading can be applied with the resulting peak stress calculated by superposition. Further, to obtain the stresses in the adhesive caused by loading the lower adherend simply interchange the subscripts 1 and 2 in the formulae. Limitations in the use of these simplified formulae are outlined in the general elastic analysis section.

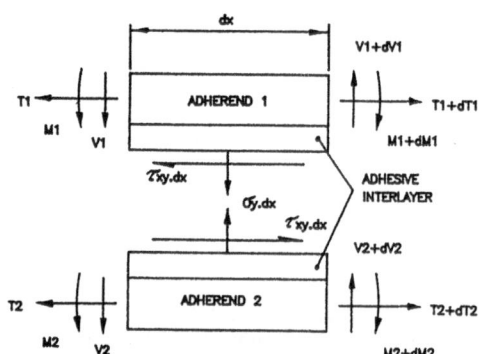

ADHEREND 1: ELASTIC MODULUS E1, POISSONS RATIO v1, THICKNESS h1

ADHEREND 2: ELASTIC MODULUS E2, POISSONS RATIO v2, THICKNESS h2

ADHESIVE : MODULUS Ea, SHEAR MODULUS Ga, THICKNESS t

Figure 2. Elemental diagram of adherend-adhesive sandwich under general loading

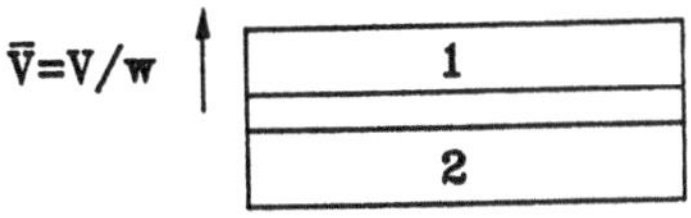

$$\sigma_V = \frac{\sqrt{2}\,\beta_1}{(\beta_1+\beta_2)^{3/4}}\,\bar{V}$$

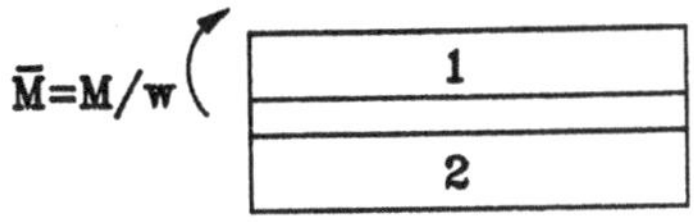

$$\sigma_M = \frac{\beta_1}{\sqrt{(\beta_1+\beta_2)}}\,\bar{M}$$

where the peel compliance factors, β_1 and β_2, are defined as:

$$\beta_1 = \frac{12\,E_a\,(1-\nu_1^2)}{t\,E_1\,h_1^3} \qquad \beta_2 = \frac{12\,E_a\,(1-\nu_2^2)}{t\,E_2\,h_2^3}$$

Figure 3. Design formulae describing peak transverse stress at left end of overlap

Elastic Simplified Shear Analysis

A similar approach to that carried out in the simplified peel analysis is used here, except only the in-plane deformation of the adhesive is considered and the adhesive peel stress is neglected as it is assumed to have a limited contribution to this in-plane deformation. Instead of adherend-bending equations adherend stress-strain equations are substituted relating adherend strains at the adhesive interface with applied loading and lastly a linear shear stress-strain relationship is assumed linking the shear stress in the adhesive to the horizontal displacements of each adherend at the adhesive interfaces by means of a shear spring analogy.

The above relations are reduced to a third-order differential equation in shear stress, shown here:

$$\frac{d^3\tau_{xy}}{dx^3} - K_1' \frac{d\tau_{xy}}{dx} = 0 \tag{3}$$

with a solution of the general form:

$$\tau_{xy} = A_2 \mathrm{Cosh} nx + B_2 \mathrm{Sinh} nx + C_2 \tag{4}$$

Boundary conditions relating shear stress derivatives with combinations of loading at either end of the overlap and equating the integral of shear stress over the overlap length with the net tensile loading on either adherend are used to find constants $A_2 - C_2$ explicitly enabling the shear stress distribution along the overlap to be described.

Simplifications similar to those made in the previous peel analysis are made here and equations are produced in two parameters describing the peak shear stress at the left end of the sandwich for separate load components. The simplified equations and the definitions of the parameters involved are given in Figure 4, below. Limitations in the use of this simplified analysis are discussed in the next section.

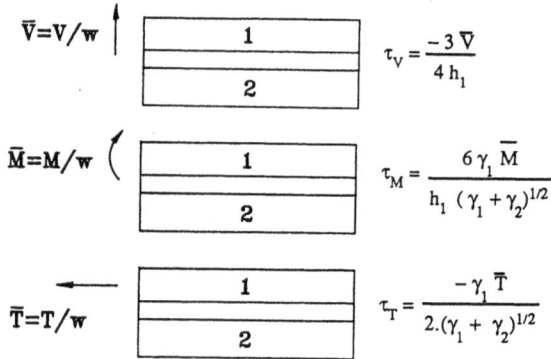

with the shear compliance factors, γ_1 and γ_2, defined as:

$$\gamma_1 = \frac{G_a}{E_1 h_1 t} \qquad \gamma_2 = \frac{G_a}{E_2 h_2 t}$$

Figure 4. Design formulae describing peak shear stress at left end of overlap

GENERAL ELASTIC ANALYSIS

Expanding the simplified analysis approach to include both components of adhesive stress the general elastic analysis is produced. This work is basically an extension of a lap joint analysis proposed by Grant [4] to allow for the general loading of the adherend-adhesive sandwich.

Considering the elemental length, dx, of overlap shown in Figure 2 an analysis is derived combining both transverse and shear stress components which were considered separately in the simplified analyses. Force and moment balance equations for both upper and lower adherends are produced. Simple bending expressions are used to relate adherend moment loading to adherend transverse displacements. At the adherend-adhesive interfaces stress-strain expressions for both adherends relate adherend loading to strain, and lastly elastic stress-strain relationships for the adhesive in both the transverse and shear directions relate adhesive stress to displacement.

The above relations are reduced to the coupled differential equations shown below:

$$\frac{d^3\tau_{xy}}{dx^3} - K_1\frac{d\tau_{xy}}{dx} = -K_2\sigma_y \tag{5}$$

$$\frac{d^4\sigma_y}{dx^4} + K_3\sigma_y = K_4\frac{d\tau_{xy}}{dx} \tag{6}$$

where K_1 - K_4 are constants describing the joint geometry and material properties.

It should be noted at this point that the differential equations shown above in (5) and (6) differ from those presented in the simplified analysis sections in that they include non-zero right hand sides, i.e. they are still coupled at an equivalent stage in their derivation. Apart from this difference they are of a similar form. The coupled effect arises purely because, contrary to the assumptions made in the simplified analyses, adhesive peel stress contributes to adherend tension and adhesive shearing contributes to the vertical deflection of the adherend. The inclusion of the coupled effect improves the accuracy of the general elastic analysis.

If the adherends are assumed to be similar then constants K_2 and K_4 reduce to zero giving right-hand sides of equations (5) and (6) equal to zero. Also constants K_1 and K_3 simplify and equate with simplified forms of the equivalent constants (K_1', K_3') shown in equations (1) and (3) for the similar adherend case. This suggests, and has been shown, that the simplified analyses are only strictly correct for balanced or similar adherends, which is nonetheless, a very common joint configuration.

To solve the coupled equations given by (5) and (6) they are expanded further to separate the variables. This process produces a seventh order differential equation in shear stress and a sixth order differential equation in transverse stress. The relevant boundary conditions relating combinations of adherend loads at the ends of the sandwich are invoked and the

thirteen simultaneous equations produced are solved using a
Gaussian elimination routine on a desktop computer. The
transverse direct and shear stress can then be fully and
accurately described along the adhesive overlap for the
elastic case.

FINITE ELEMENT ANALYSIS COMPARISON

To validate the general elastic analysis, and also the general
non-linear analysis introduced later in this paper, a finite
element (FE) analysis of an adherend-adhesive sandwich was
carried out. A mesh was constructed using 280 8-noded
isoparametric plane strain finite elements to represent an
epoxy-aluminium joint with thicknesses 0.25 and 1.00mm
respectively and with an overlap length of 12.5mm. Figure 5
shows the finite element mesh with details of the left and
right ends of the overlap region showing load application and
constraint methods respectively. Note from the figure that
five elements were used to model the thickness of the adherend
and four elements were used across the adhesive thickness. A
small displacement option in the finite element analysis was
used to match the assumptions made in the analysis being
validated.

The load case chosen for the elastic comparison was a single-
lap type loading with load values per unit width given by:

Tension loading $\quad T_{11} = T_{22} = 725$ N/mm
$\qquad T_{12} = T_{21} = 0$ N/mm
Shear loading $\qquad V_{11} = V_{22} = -40$ N/mm
$\qquad V_{12} = V_{21} = 0$ N/mm
Moment Loading $\quad M_{11} = -M_{22} = -200$ Nmm/mm
$\qquad M_{12} = M_{21} = 0$ Nmm/mm

Note the sign convention shown in Figure 1a applies in this
case. Load values T11, and M11 are applied directly to the
left end of the upper adherend as a linear distribution of
stress defined by concentrated loads or forces applied
directly to the nodes. Figure 5 shows the distribution when
these loads are applied together. A parabolic distribution
was used to apply the shear loading in a similar manner. As
the mesh is constrained across the right hand face of the
lower adherend by restraining the nodes to have zero
displacement in both the x- and y-directions the finite
element analysis calculates reactions internally to balance
the joint in a similar way to the general elastic analysis.

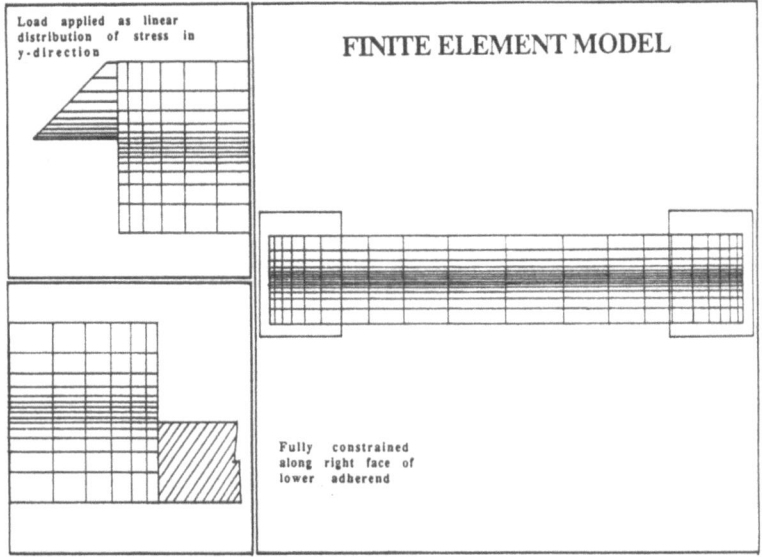

Figure 5. Finite element model

Figure 6 shows the transverse direct and shear stress distributions obtained from the above load case plotted along the adhesive overlap region when the following material properties were used:

Adhesive	Elastic Young's Modulus	= 2500 N/mm^2
	Elastic Shear Modulus	= 892.9 N/mm^2
Adherends	Elastic Young's Modulus	= 70000 N/mm^2
	Poisson's ratio	= 0.3

We note from both stress curves shown in Figure 6 that there is very good agreement between the general elastic analysis, denoted by the continuous line, and the FEM, denoted by hexagons. The distributions themselves are well reported and little discussion is necessary, however, an interesting point to note is the way the FEM attempts to model the stress-free state for shear stress at the ends of the overlap. The mesh chosen is not fine enough to model this effectively but the much lower values of shear stress at the overlap ends show the effect, it also shows how very localised this effect is.

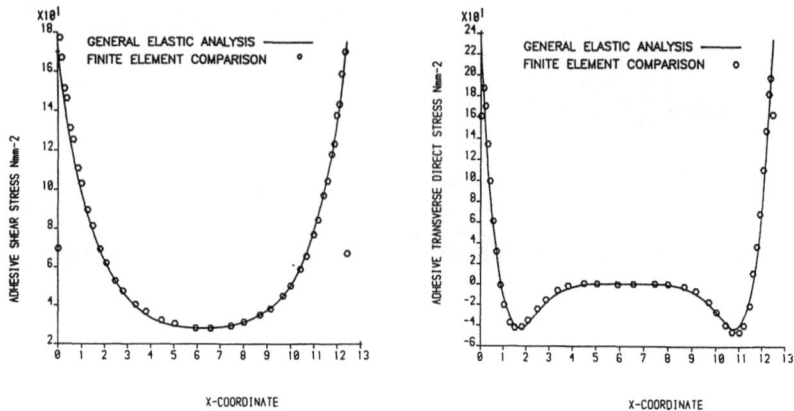

Figure 6. Stress distributions from the general elastic analysis and the finite element comparison

THE NON-LINEAR ADHESIVE JOINT PROBLEM

To accurately quantify failure, joint strength prediction work requires the inclusion of material non-linearity into the analysis formulation. Also the inclusion of non-linearity has been shown to reverse trends illustrated by elastic analysis, such as adhesive thickness effects. Bonded joint analyses in which material non-linearity is partially treated have been in existence for some time. The following two sections briefly outline two of these analyses and their solution methods. The third section introduces a general non-linear analysis which extends the non-linearity to fully model adhesive material non-linearity. Finite element validation work and comparisons between the general non-linear analysis and others are shown and discussed.

GRANT'S APPROACH TO PLASTICITY

Grant's approach to the single-lap joint problem [4] was to introduce adhesive non-linear material behaviour into Volkersen's shear-lag analysis [1]. The analysis only includes the shear stress and strain caused by tension loading of the adherends and neglects both the shear and moment load components and the transverse stress resulting from them. The Engineering Science Data Unit (ESDU) market the analysis as a non-linear lap joint analysis [8]. The program is not strictly applicable to single-lap configurations because the bending effects are ignored but can be used to good effect to analyse double-lap joints where the transverse stress component is largely suppressed.

Volkersen's simple differential equations which describe the
adhesive shear strain in terms of the differential stretching
of the adherends are obtained from the elemental diagram shown
in Figure 7.

The shear strain, γ, is expressed as a function of the
relative displacements of adherends 1 and 2, given by u_1 and
u_2 respectively.

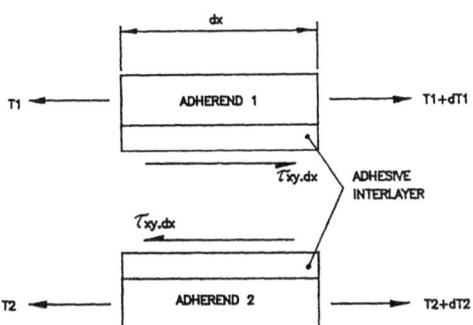

Figure 7. Grant's approach to plasticity

The equations for horizontal equilibrium and the derivative of
the adhesive shear stress are reduced to finite-difference
form for the upper adherend by introducing a horizontal force
balance for the joint and eliminating the tension in the lower
adherend, T_2. The finite difference equations have the
following form:

$$dT_1 = -\tau_{xy}\, dx \tag{7}$$

$$du = \left\{ \frac{T}{t_2 E_2} - T_1 \left(\frac{1}{t_1 E_1} + \frac{1}{t_2 E_2} \right) \right\} dx \tag{8}$$

where

$$u = u_2 - u_1 = t_a \gamma \tag{9}$$

The overlap length is split into n equal increments of length
dx and the equations are solved using a marching technique.
An initial shear strain is assumed at the left end of the
overlap and from a simple linear-reciprocal approximation to

the stress-strain curve of the adhesive a value of shear
stress corresponding to the initial guess of shear strain is
found. The finite-difference equations (13) and (14) are then
used to predict new values of T_1 and u (and hence shear
strain, γ), at the end of the first increment of length. The
procedure is repeated until the end of the joint is reached,
where the boundary condition for the single lap joint states
that T_1 should equal zero. The tension calculated in the
adherend at the right end of the overlap is compared with zero
and the initial guess of shear strain at the left end of the
overlap is altered. The finite-difference procedure is
repeated until $T_1 \cong 0$ at the right end of the overlap, when
the shear stress and strain distributions can be fully
described in the adhesive layer. A problem of instability for
joints with longer overlap lengths and corresponding areas of
low almost constant stress was noted by Grant. A refinement
to the analysis, involving a slightly more complicated finite-
difference procedure, was introduced, allowing solutions for
these longer joints to be found. Results from Grant's
analysis are shown in Figure 16, where they are compared with
results from other plastic analyses.

HART-SMITH'S APPROACH TO PLASTICITY

Hart-Smith [3], considers the balanced single-lap joint as
shown in Figure 8, below.

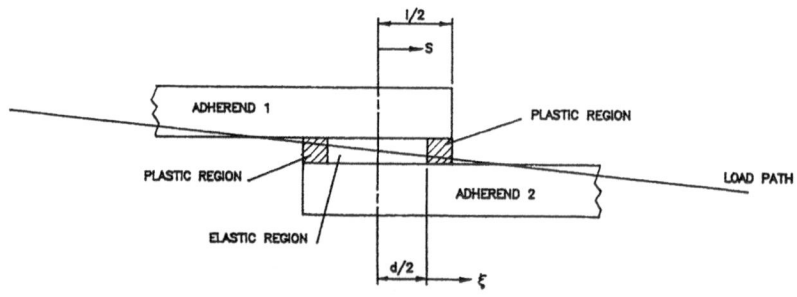

Figure 8. Hart-Smith's approach to plasticity showing
his balanced lap joint with co-ordinate
definitions

The transverse stress is treated elastically and the shear
stress as elastic-perfectly plastic. The adhesive is split
into three regions: a central elastic region and two outer

plastic regions. Co-ordinates s and ξ are defined in these regions as shown in Figure 8. The shear stress in the elastic region is represented by the third order differential equation shown in equation (10) below, with a solution of the general form given by equation (11).

$$\frac{d^3 \tau_{xy}}{ds^3} - 4(\lambda')^2 \frac{d\tau_{xy}}{ds} = 0 \tag{10}$$

$$\tau_{xy} = A_3 \text{Cosh}(2\lambda' s) + B_3 \text{Sinh}(2\lambda' s) + C_3 \tag{11}$$

Note that equation (10) is the same as the corresponding shear stress equation in the general elastic analysis given by (5).

The plastic regions in which the shear stress is assumed to be a constant plastic value of τ_p and hence $d\tau_{xy}/d\xi = 0$, has a shear strain distribution represented by this equation:

$$\frac{d^3 \gamma}{d\xi^3} = 0 \tag{12}$$

with a solution of the general form:

$$\gamma = A_4 \xi^2 + B_4 \xi + C_4 \tag{13}$$

For a balanced joint with a corresponding symmetrical shear stress distribution B_3 can be assumed equal to zero and hence Hart-Smith solves his equations for the unknown constants A_3, C_3, $A_4 - C_4$ and the unknown ratio of the length of elastic and plastic regions, d/l. The six boundary conditions used relate the shear strain and the first and second derivatives of shear strain at the elastic-plastic region interface, the adhesive shear strain to the strain in the adherend at the adherend-adhesive interface at the end of the overlap and the integral of shear stress over half the overlap length with one half of the applied load.

The equations are solved for the six unknowns and the shear stress and strain distributions for the elastic and plastic regions in the overlap can then be fully described. Figure 16 shows maximum strain values for Hart-Smith's analysis and contains comparisons with other plastic analyses.

GENERAL NON-LINEAR ANALYSIS

As with the general elastic approach an attempt has been made to produce an analysis that is applicable to a wide range of joint configurations but to include non-linear material characteristics in it's formulation.

Again, the same elemental diagram used in the general elastic analysis as shown in Figure 2 is considered, but the interlayer is assumed to be non-linear, with each stress component a non-linear function of all strain components in the adhesive. The functions are described in the deformation theory of plasticity [9]. This assumes a direct relation between stress and strain in the plastic regime and does not rely on an incremental loading approach and is valid if increasing loads are applied in proportion. A system of equations defines the deformation theory and relates the non-linear stress and strain with two parameters, defined the secant modulus, E_s, and the plastic Poisson's ratio, v_p, which are given by:

$$E_s = \frac{\sigma}{\varepsilon} \tag{14}$$

$$v_p = \frac{1}{2} \left\{ 1 - \frac{E_s}{E} (1 - 2v) \right\} \tag{15}$$

The secant modulus represents the slope of the straight line from the origin to any point on the uniaxial stress-strain curve, and is shown diagrammatically below in Figure 9. The plastic Poisson's ratio varies between its elastic value, v, and an asymptote of 0.5 for the fully plastic case.

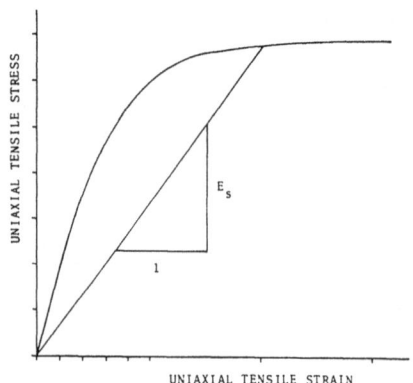

Figure 9. Definition of secant modulus

The analysis derivation is approached in a similar way to the general elastic analysis with a few extensions. Force and moment equilibrium expressions relating the adherend loading to the stresses in the adhesive are defined as in the previous analysis. The stresses are then substituted by non-linear functions of strain given by deformation theory, resulting in equations (16) - (18) below. Expressions relating the shear strain derivative in the adhesive to the difference in adherend surface strains is also used, equation (19), and a final expression relating the second derivative of transverse strain in the adhesive to the applied bending load is also used. The above expressions are shown below represented by six first-order non-linear differential equations:

$$\frac{dT_1}{dx} = \tau_{xy} = \frac{E_s \, \gamma_{xy}}{2(1 + \upsilon_p)} \tag{16}$$

$$\frac{dV_1}{dx} = \sigma_y = \frac{E_s \, \varepsilon_y}{(1 - \upsilon_p^2)} \tag{17}$$

$$\frac{dM_1}{dx} = V_1 - \left(\frac{h_1 + t}{2}\right) \tau_{xy} = V_1 - \left(\frac{h_1 + t}{4}\right) \frac{E_s \gamma_{xy}}{(1 + \upsilon_p)} \tag{18}$$

$$\frac{d\gamma_{xy}}{dx} = \frac{1}{t} \left\{ \frac{1}{E_1 h_1} \left(T_1 - \frac{6M_1}{h_1}\right) - \frac{1}{E_2 h_2} \left(T_2 + \frac{6M_2}{h_2}\right) \right\} \tag{19}$$

$$\frac{dK}{dx} = \frac{12}{t} \left\{ \frac{(1 - \upsilon_2^2) \, M_2}{E_2 h_2^3} - \frac{(1 - \upsilon_1^2) \, M_1}{E_1 h_1^3} \right\} \tag{20}$$

$$\frac{d\varepsilon_y}{dx} = K \tag{21}$$

An extra variable is introduced, K, defined as the first derivative of transverse strain to allow the second derivative of transverse strain to be represented by two first order equations. Variables T_2 and M_2, tensile and moment loading in the lower adherend are substituted for by corresponding upper adherend values (T1, M1) and end loads by considering equilibrium of a length, x, of the adherend-adhesive sandwich.

where T_{11}, T_{21}, M_{11} and M_{21} are the boundary condition loads at the left end of the overlap and T_1, M_1, T_2 and M_2 are the

178

tensile and moment loads in the upper and lower adherends respectively at a distance, x, from the left end of the overlap.

A von Mises yield criterion [10] is used to calculate an equivalent strain, $\bar{\varepsilon}$, on a uniaxial stress-strain curve as a function of transverse and shear strain found by substituting for ε_y and γ_{xy} in the following expression, derived by substituting for stresses in the von Mises yield function.

$$\bar{\varepsilon} = \frac{1}{(1 - \upsilon_p^2)} \left\{ (1-\upsilon_p+\upsilon_p^2)\varepsilon_y^2 + \frac{3}{4}(1-\upsilon_p)^2\gamma_{xy}^2 \right\}^{1/2}$$

(22)

The above six differential equations (16) - (21), the yield criteria (22) and any continuous mathematical function used to represent the uniaxial stress-strain curve of the adhesive are solved using a boundary-value finite-difference solution algorithm [11]. The solution algorithm is shown schematically in Figure 10.

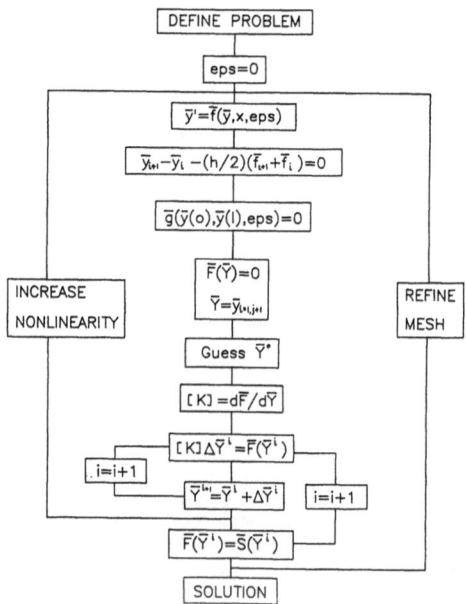

Figure 10. Schematic of finite-difference solution method

A variable, eps, termed the non-linearity parameter is introduced into the six differential equations allowing the problem to be increased in complexity from linear (eps = 0) to fully non-linear (eps = 1.0). In this way, the amount of non-linearity in the problem can be increased gradually by incrementing eps, the solution to one problem forming an approximate initial guess to the subsequent, more complex solution. The six equations (16) - (21) are each written in finite-difference form using a first-order approximation. for the derivatives, these are represented as f() in the schematic diagram. The overlap region is split into n equal increments of length, h, the finite-difference equations f() are specified on each region giving 6n equations in 6(n+1) unknowns, the values of the variables at each mesh point. Boundary conditions, g, are invoked defining load values at either end of the sandwich providing the remaining 6 equations required to enable solution. The finite-difference equations and the boundary conditions together form a set of non-linear equations F().

An initial guess for the variables at each mesh point, Y'(), is made. This can be user supplied or provided by the algorithm by linearly interpolating between non-zero boundary conditions. A Jacobian [K] is determined representing the derivative of F() with respect to Y() and a Newton iteration routine is used to improve the initial guess. After one Newton iteration the level of non-linearity is increased and the equations recomputed, this procedure is repeated until the full non-linear equations are represented.

When the full non-linear case is achieved a correction for the Newton iteration method can be obtained by including higher order terms in the finite-difference equations. In this manner it is attempted to reduce all errors at all points in the mesh to within a specified tolerance band. If the tolerance is reached the solution is output, if not the mesh is refined locally around the points with the largest errors, the Jacobian is recomputed and the Newton iteration with correction is carried out until a solution is found. The method of solution described has been used successfully with a large range of joint configurations and load combinations.

FINITE ELEMENT COMPARISON

To validate the non-linear analysis described above a non-linear finite element analysis was carried out using the model described earlier in the paper, and shown in Figure 5. Hart-Smith single-lap joint loading [3] was assumed and applied in a similar way to that described in an earlier section.

The non-zero load values applied were as follows:

Tensile loading T_{11} = 400.0 N/mm
Shear loading V_{11} = -29.58 N/mm
Moment loading M_{11} = -65.0 Nmm/mm

Again the reactions are provided internally by the finite element program and produce the balanced loading required in a classic lap joint problem.

Non-linear adhesive material properties were assumed. The general non-linear analysis, in this case, used a hyperbolic-tangent approximation to the uniaxial stress-strain curve of the following form:

$$\bar{\sigma} = E_a \text{Tanh} \left(\frac{A\bar{\varepsilon}}{E_a} \right)$$

(23)

with: E_a = Elastic Young's Modulus = 1875.0 N/mm²
        A = Asymptotic maximum stress value = 69.28 N/mm²
        σ, ε = Equivalent stress and strain respectively

Equation (23) is shown graphically in Figure 11, as is the bi-linear approximation to the stress-strain curve used in BICEPS-LOCO and Hart-Smith and the linear-reciprocal approximation used in Grant's analysis.

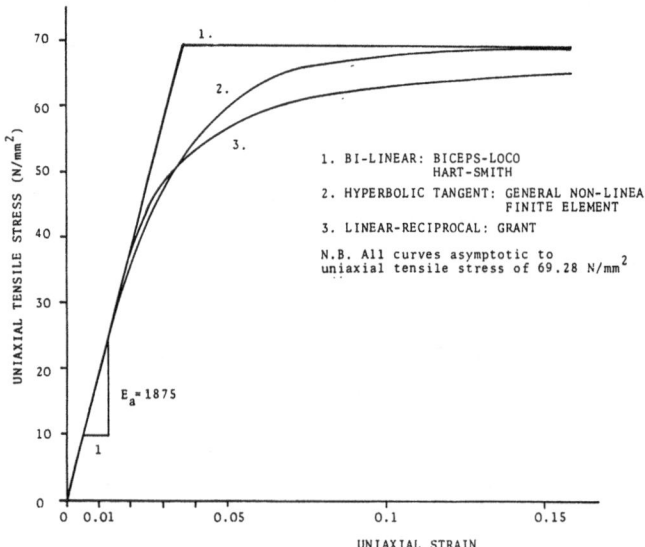

Figure 11.   Approximations to stress-strain curve used in non-linear analyses.

The finite element program used a cubic spline approximation to the stress-strain curve with three internal knots, .which models the curve shown in Figure 11 almost exactly.

The total level of loading was applied in thirty-seven load fractions, the first fraction, enough to cause initial yielding in the adhesive layer, was 0.0689 of the total applied load, with the remaining load being applied in thirty-six equal increments. The finite element model converged easily in one or two iterations for each increment of load applied.

Finite element results averaged across the adhesive layer for transverse direct stress and strain and shear stress and strain are compared with results from the general non-linear analysis, using the .same load case, in Figures 12 and 13 respectively.

Referring to the transverse stress and strain details shown in Figure 12 we can see fairly close agreement between the finite element results, denoted by triangles, and the general non-linear analysis, denoted by the continuous line. The agreement between the finite element and the general non-linear analysis results for the non-linear case is not as close as for the linear case, but is still well within an acceptable error band. The shear stress and strain results shown in Figure 13 agree very closely with each other. An interesting point to note is the downturn of the end of the overlap in the shear stress graph. This effect is due to the inclusion of the transverse strain in the expression for the yield criterion which, increasing at a faster rate than the shear, causes the location of the yield surface to move towards the transverse stress, any increase in yield due to hardening is more than absorbed by a corresponding larger increase in the peel stress. This effect has not been previously reported to the best of the authors' knowledge, and it is pleasing to note the close agreement of the finite element results with the non-linear analysis results particularly in this region.

The general non-linear analysis will provide an extremely useful tool if joint strength can be obtained in terms of a general level of strain in the adhesive.

To fully appreciate the effect of including material non-linearity into the analysis formulation it is necessary to compare elastic and plastic analyses. Figures 14 and 15 show these comparisons and it can clearly be seen how the elastic analysis underestimates the strains while overestimating the stresses. The load case displayed in these figures is of a similar type to that above, but at about twice the total load and sufficient to put the whole overlap region well into yield.

As a final comparison, results from the general non-linear analysis have been compared with three other analyses which

model adhesive non-linearity to some degree. The analyses
used are Grant's shear-lag analysis [4] and Hart-Smith's lap
joint analysis [3], both described earlier in this paper, and
BICEPS-LOCO [12] a non-linear adhesive property program
developed at Harwell Laboratory. Also plotted are the results
from the non-linear finite element analysis.

Referring to Figure 16 containing the shear stress and strain
results for all of the above analyses it is noted that Grant's
shear-lag analysis shows fair correlation with the finite
element results. The results obtained for Grant's analysis
are reasonable, even though the analysis neglects adherend
bending, this is because the load case chosen is predominantly
tension and Grant models this well. The shear results for a
load case with a larger degree of bending in the adherends
would not agree as well.

Hart-Smith's single-lap joint analysis has been incorporated
by using the comprehensive parametric studies for the lap
joint configuration given in his report. It is possible to
calculate maximum strain values from these studies, and these
agree well with finite element values, but it should be noted
that the transverse stress and strain which is considered
elastically in this analysis will be in error.

BICEPS-LOCO is compared in both the shear and transverse
directions (Figure 17) and is shown to agree very well with
both the FEM and the general non-linear analysis in the
transverse direction for both stress and strain. The shear
stress and strain results from BICEPS-LOCO, however, differ
from those predicted by finite-element. After consulting the
authors it would seem likely that this is because the Harwell
program incorporates a hydrostatic sensitive yield criterion,
a feature that was not included in these initial tests. The
authors have noted similar effects when hydrostatic criteria
are applied in their analysis and this will be reported more
fully in a later paper.

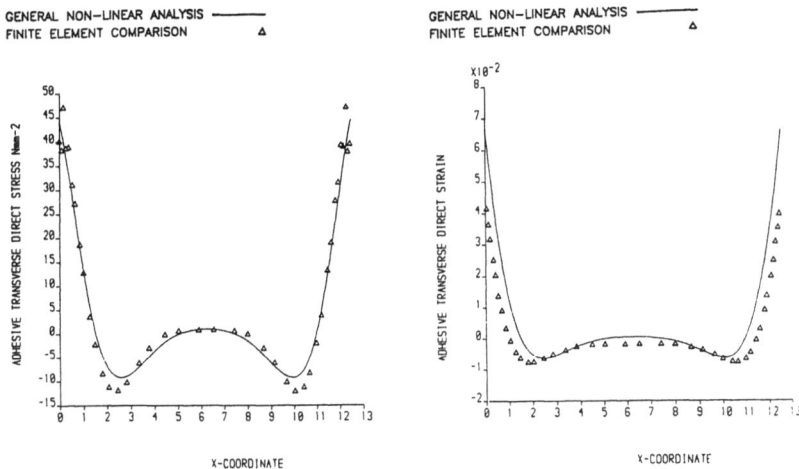

Figure 12. Transverse stress and strain distributions along overlap. General non-linear analysis compared with non-linear finite element analysis.

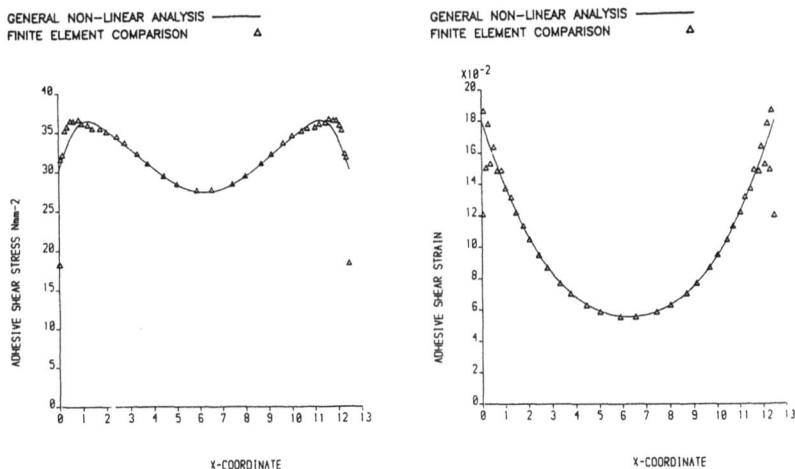

Figure 13. Shear stress and strain distributions along overlap. general non-linear analysis compared with non-linear finite element analysis.

184

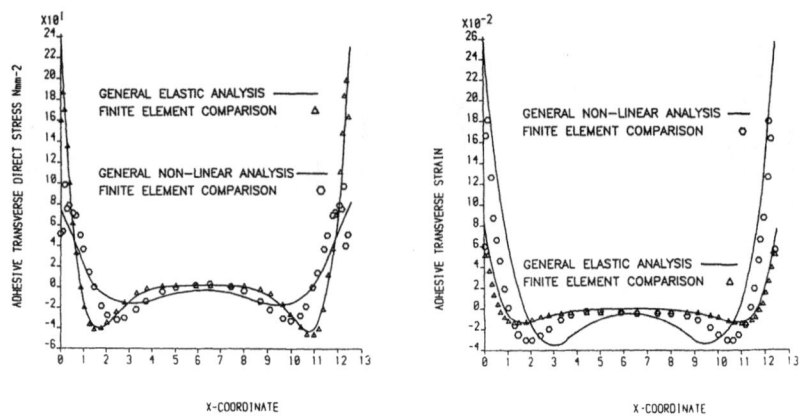

Figure 14. Transverse stress and strain distributions along overlap. Elastic and plastic comparisons showing general and finite element analyses.

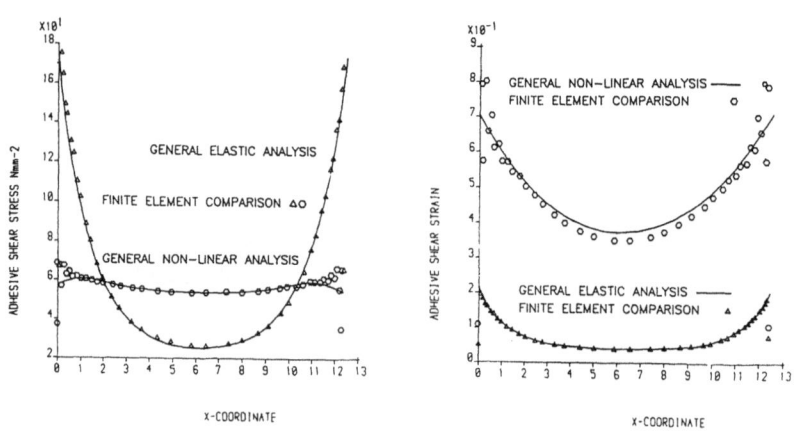

Figure 15. Shear stress and strain distributions along overlap. Elastic and plastic comparisons showing general and finite element analyses.

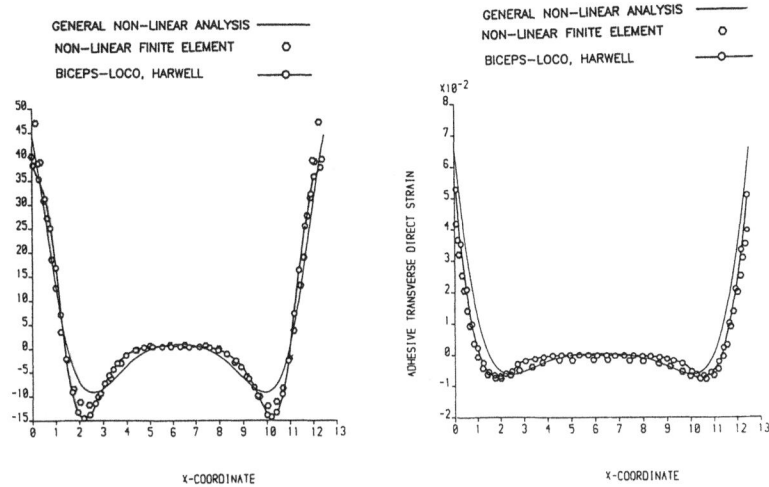

Figure 16. Transverse stress and strain distributions along overlap. Various plastic analyses.

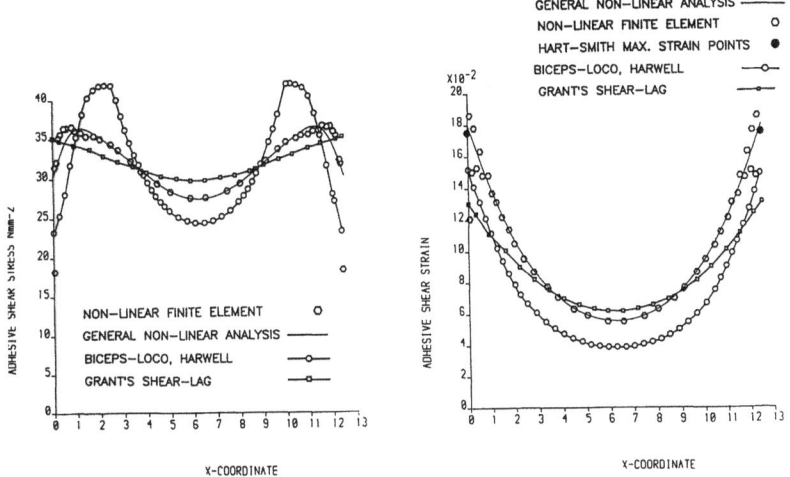

Figure 17. Shear stress and strain distributions along overlap. Various plastic analyses.

CONCLUSIONS

The following conclusions are drawn from the work presented here:

1.   The general elastic analysis gives good results for the elastic configuration. its main advantage over similar analyses being its ability to model general joint configurations. The analysis requires implementation on a microcomputer to obtain stress results.

2.   Balanced adherend stresses can be found using explicit constants on a calculator using the simplified elastic analysis. The procedure involved is a little tedious.

3.   Peak transverse and shear stresses are found using simple two parameter formulae. They are found to be exact for balanced joints and provide a simple initial estimate of joint strength, especially useful at a preliminary design stage.

4.   Various approaches to the solution of the non-linear problem are outlined, all with different assumptions and degrees of non-linearity. These have been shown to produce results with corresponding degrees of accuracy.

5.   The approach introduced in the general non-linear analysis is an extension of the general elastic analysis, and is different from the other approaches mentioned in the following respects: the results are continuous over the overlap length which is not split into separate elastic and plastic regions; a continuous model for the stress-strain curve can be used allowing for greater accuracy in its modelling; the analysis is useful for considering general joint configurations and not just balanced joints.

6.   The general non-linear analysis agrees well with non-linear finite element analysis.

REFERENCES

1.   Volkersen, O., **Rivet Strength Distribution in Tensile-Stressed Rivet Joints with Constant Butt Strap Cross Section**, Luftfahrforschung, Vol. 15, No. 1/2, p. 41, 1938.

2.   Goland, M. and Reissner, E., **The Stresses in Cemented Joints**, Journal of Applied Mechanics, Trans ASME 66, p. A17, 1944.

3. Hart-Smith, L.J., **Adhesive Bonded Single-Lap Joints - Technical Report**, Contract No. NASA-CR-112236, Douglas Aircraft Co., January 1973.

4. Grant, P. and Taig, I.C., **Strength and Stress Analysis of Bonded Joints**, B.Ae. Report SOR(P) 109, Warton Division, 1976.

5. Kaeble, D.H., **Theory and Analysis of Peel Adhesion: Bond Stresses and Distributions**, Trans Soc of Rheology IV, p. 45, 1960.

6. Hart-Smith, L.J., **Adhesive Bonded Scarf and Stepped-Lap Joints**, NASA CR-112237, McDonnel-Douglas Corp. January 1973.

7. Hart-Smith, L.J., **Adhesive Bonded Double-Lap Joints**, NASA CR-112235, McDonnel-Douglas Corp, January 1973.

8. Engineering Sciences Data Unit, Item No. 79016, **Inelastic Shear Stresses and Strains in the Adhesives Bonding Lap Joints Loaded in Tension or Shear**, September 1979.

9. Jensen, W.R., Falby W.E., and Prince N., **Matrix Analysis Methods for Anisotropic Inelastic Structures**, Technical Report: AFFDL-TR-65-770, 1966.

10. Mises, R. von, **Mechanik der festen körper im plastisch-deformablen Zustand**, Gottingen Nachrichten Maths-Phys, p. 582, 1913.

11. **NAG Fortran-Library Mark 11, Chapter D02 - Ordinary Differential Equations - (D02RAF)**, National Algorithms Group, January 1984.

12. Livey, D.T., and McCarthy, J.C., **Composite/Metal Jointing Technology for Vehicle Weight Reduction**, Contract No. EEC-4-261 UK Summary Report, AERE Harwell, Oxon, UK.

12

FINITE ELEMENT MODELLING APPLIED TO CRACK PROPAGATION STUDIES IN BONDED COMPONENTS

J.S. Crompton and J.D. Clark
Alcan International Limited,
Southam Road,
Banbury,
Oxon,
OX16 7SP.

INTRODUCTION

The use of adhesively bonded aluminium for structural applications can lead to significant benefits in weight saving, structural stiffness and ease of fabrication. The bonded joints must be able to tolerate extremes of loading and environment. Within a structure, a bond may be subjected to combinations of both static and cyclic loads during service. In the case of cyclic loading, satisfactory bond performance can be judged to occur when fatigue damage is apparent in the metal before the adhesive. A judgement of this nature requires a knowledge of the cyclic loading characteristics of the adhesive used to form the bond.

In most cases the process of fatigue failure can be viewed as one comprising a crack initiation stage followed by a crack propagation stage leading to ultimate failure. In practice, the time for a crack to initiate within a joint may be small due to the complex stress distributions present in joints and the distribution of inhomogeneities found within bonds. Thus, the critical stage in the lifetime of a bonded joint is that associated with the growth of the embryonic flaw to a size at which catastrophic failure of the joint occurs.

The majority of studies of bond failure have been associated with the point of final instability of the joint (1,2). Several studies have, however, been conducted into the fatigue crack propagation rates of bonded or composite components (3,4). To investigate fully the extent to which

188

this behaviour is important in determining the lifetime of a structure it
is necessary to examine a range of crack propagation rates. At high crack
propagation rates ($> 10^{-3}$ mm/cycle) structural instability results. At
low crack propagation rates ($< 10^{-7}$ mm/cycle) loading conditions may exist
such that negligible crack advance occurs. This latter behavioural regime
allows threshold operating loads to be defined such that pre-existing
defects are non-propagating. The establishment of damage tolerant design
procedures of this type requires that crack propagation rates for
different joint configurations, loading conditions, bond line thicknesses
and crack lengths, can be predicted. To obtain this appropriate
parameters are required which accurately correlate crack propagation data
under a range of operating conditions. It is the purpose of the present
work to examine the fatigue behaviour of a bonded aluminium component and
provide details of the analytical procedures used in evaluating the crack
propagation data.

EXPERIMENTAL PROCEDURE

Fatigue crack propagation studies were conducted on a modified
compact tension specimen (Figure 1) consisting of two aluminium blocks
bonded with a high strength single part epoxy adhesive. Two bond line
thicknesses were investigated namely 0.5 mm and 2.0 mm. Loading was
conducted using a servo-hydraulic system operating at a frequency of 10 Hz
and with the ratio of minimum to maximum load equal to 0.1.

Specimens were notched in the centre of the bond line to a nominal
a/W of 0.3. During loading, crack extension was monitored by conductive
gauges bonded to the side of the test specimen. Crack growth was obtained
under conditions of variable load control according to:

$$\Delta P = \Delta P_s \exp (-Cda)$$

where ΔP = load range,
   ΔP_s = starting load range
   C = rate constant
   da = crack length increment

Testing was normally conducted using a value of $C = \pm 0.2 \text{ mm}^{-1}$.

This method of testing enabled a wide range of crack growth rates to
be examined between 10^{-3} mm/cycle and 10^{-7} mm/cycle. Threshold behaviour

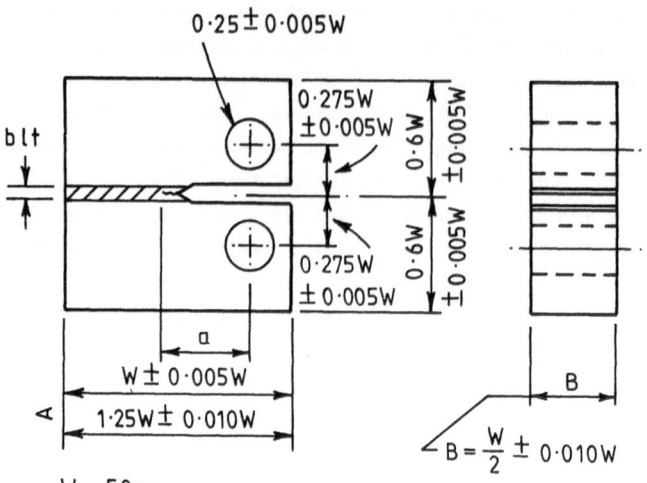

0·25±0·005W

0·275W ±0·005W

0·6W ±0·005W

0·275W ±0·005W

0·6W ±0·005W

blt

a

W ± 0·005W

1·25W ± 0·010W

A

B

$B = \dfrac{W}{2} \pm 0·010W$

W = 50mm.

BOND LINE THICKNESS = 2mm. or 0·5mm.

FIGURE 1. Bonded aluminium compact tension specimen

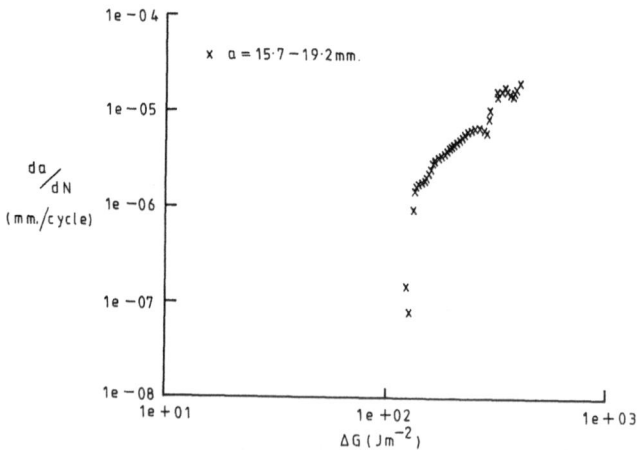

x a = 15·7 − 19·2mm.

$\dfrac{da}{dN}$
(mm./cycle)

ΔG (Jm^{-2})

FIGURE 2. Fatigue behaviour of 2mm. bondline specimen

was considered to be that associated with growth rates $< 10^{-7}$ mm/cycle.
By suitable choice of operating conditions many evaluations of the crack
growth rate behaviour with differing loads were obtained from one
specimen, thus providing data from adhesive having common
characteristics.

Results and Discussion

Previous work in fatigue crack propagation of composite bonded
components has used the cyclic strain energy release rate, $\Delta G = Gmax-Gmin$,
to correlate the crack propagation data. To analyse the data in this form
requires that calibration expressions are available for different specimen
geometries relating the applied load and crack length to the strain energy
release rate, G. A value of ΔG can be calculated for a given specimen
geometry using the relationship

$$\Delta G = \frac{\Delta P^2}{B} \left(\frac{\delta c}{\delta a}\right)$$

where ΔP = applied load range, B = specimen thickness and $\left(\frac{\delta c}{\delta a}\right)$ is the rate
of change of specimen compliance with crack length. For the current
specimen geometry the form of $(\delta c/\delta a)$ was determined experimentally for
specimens a 0.5 mm and 2.0 mm bond line thickness. The values of ΔG for
the bonded joints can then be determined for specific values of applied
load range and crack length. Thus, the crack growth rate data can be
related to the values of ΔG applied to the specimen; an example of the
relationship between applied ΔG and crack growth rate da/dN is given in
Figure 2 for data obtained on a specimen containing a 2 mm bond line
thickness. Similar behaviour was observed in specimens containing 0.5 mm
bond line thicknesses.

With decreasing ΔG it can be seen that there is a corresponding drop
in the crack growth rates. At $\Delta G \simeq 110$ Jm^{-2} a transition in behaviour is
evident with a sharp drop in crack growth rates observed on decreasing ΔG.
This type of behaviour has been interpreted (5) as indicating a threshold
type response thus providing conditions under which negligible crack
growth occurs.

To examine the generalised applicability of ΔG to providing a
correlation of crack growth data, several decreasing load tests were
conducted on the same specimen. In this way the growth rates of cracks in
the epoxy resin were obtained from different absolute crack lengths.

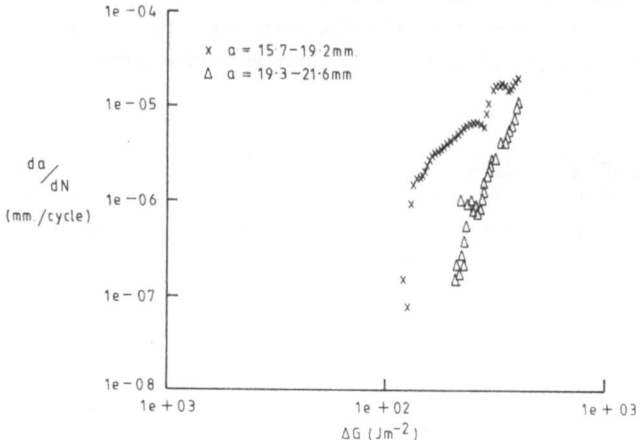

FIGURE 3. Fatigue behaviour of 2mm. bondline specimen

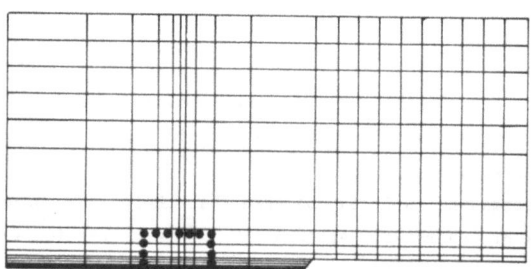

•••• typical area submodelled

FIGURE 4. Coarse model of compact tension specimen

Figure 3 presents the crack growth rates plotted against ΔG for two runs on the same specimen in which run 1 was obtained for $15.73 < a < 19.2$ and run 2 was for $19.3 < a < 22$. It is readily apparent that there is little agreement between the two sets of data. At a selected value of ΔG crack growth rates may vary by at least an order of magnitude. The exact reason for this discrepancy is not readily apparent but does indicate that ΔG does not provide a unique parameter against which crack growth rates can be correlated.

The strain energy release rate, G, is a measure of the overall energy required to provide crack extension. As such it can be related to the mean crack tip stresses only in idealised situations. Since these crack tip stresses govern local material failure, and thereby crack advance, further work examined the use of crack tip stress intensity solutions to correlating crack growth behaviour.

Crack tip stress intensities were obtained from elastic finite element modelling using ANSYS. The specimen was modelled using a cut boundary displacement method (6) in which the modelled displacements from the coarse model of the whole specimen are applied to a fine scale sub-model surrounding the crack tip. The elements around the crack tip are hybrid 3D elements developed (7) to provide information at crack tips. Details of the coarse model are presented in Figure 4 and the sub-model in Figure 5. The elastic material constants used for the calculations are given in Table 1. Output from the model was used to provide stresses

TABLE 1
Elastic Constants

	Young's Modulus, GPa	Poisson's Ratio
Aluminium	69.0	0.33
Epoxy	3.9	0.33

within the specimen (Figure 6) and crack tip stress intensity values, K, are obtained using the method developed by Tracey (7). These values were subsequently checked for convergence and reproducibility.

By obtaining K for values of (a/W) between 0.3 and 0.85, calibration equations for the different specimen geometries were developed. Figure 7

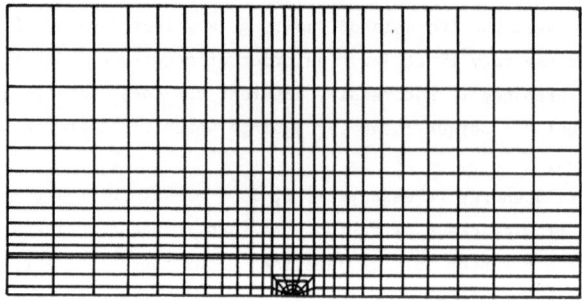

FIGURE 5. Submodel of compact tension specimen

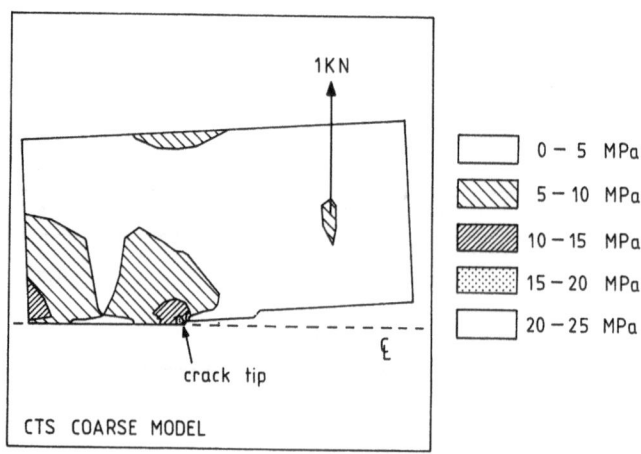

FIGURE 6. Finite element model of upper half of CT specimen showing contours of deviatonic stress

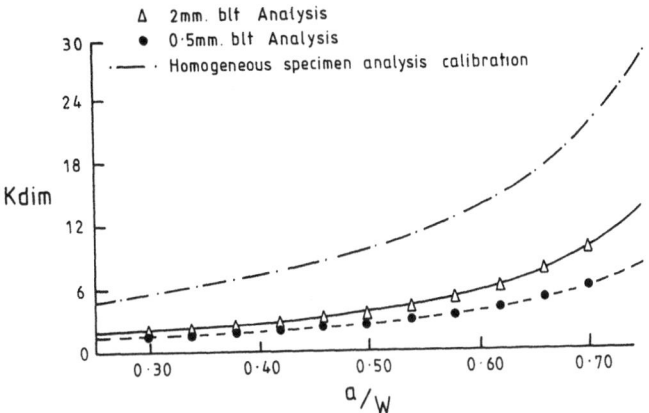

FIGURE 7.   Finite element stress intensity calibrations
for bonded CT specimens

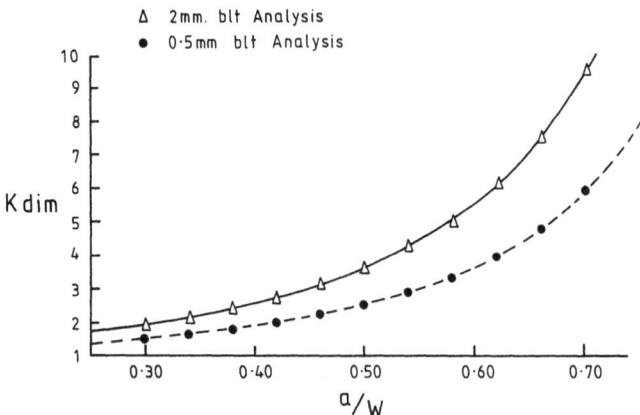

FIGURE 8.   Finite element stress intensity calibrations
for bonded CT specimens

shows the non-dimensional form of K versus (a/W) for homogeneous material; these results are in complete agreement with those generated by other workers (8). The results for specimens with 2 mm and 0.5 mm bond line thicknesses are presented in Figure 8. The analytical data were used to develop equations relating K to (a/W) using a least squares fitting routine. The equations developed are given in Table 2.

TABLE 2
Stress Intensity Calibration Equations for Bonded
Compact Tension Specimens

---

2 mm bond thickness

$$K = \frac{P}{BW^{\frac{1}{2}}} \frac{(2 + a/w)}{(1 - a/w)^{3/2}} \left[0.48 + 0.227\ (a/w) - 1.028\ (a/w)^2 + 1.638\ (a/w)^3 - 0.509\ (a/w)^4 \right]$$

0.5 mm bond thickness

$$K = \frac{P}{BW^{\frac{1}{2}}} \frac{(2 + a/w)}{(1 - a/w)^{3/2}} \left[0.396 + 0.158\ (a/w) - 1.15\ (a/w)^2 2 + 1.73\ (a/w)^3 - 0.736\ (a/w)^4 \right]$$

From these analyses it is apparent that the crack tip stress intensities in a bonded component are reduced compared to those found in a homogeneous specimen. Similar behaviour has been reported in the analytical solutions of Erdogan and Gupta (9) and Hilton and Sih (10). In addition, reducing the bond line thickness from 2 mm to 0.5 mm effects a reduction in the crack tip stress intensities.

Using the calibration equations given in Table 2, the crack propagation data were re-evaluated using the cyclic stress intensity as the correlation parameter. For a component having a 0.5 mm bond line thickness the crack growth rates for two runs on the same specimen are given in Figure 9. It can be seen that, in this case, good agreement is obtained between the crack growth rates measured at different absolute crack length. Thus, ΔK represents a parameter capable of providing correlation of crack propagation data.

The crack growth response of a specimen containing a 2 mm bond line thickness is presented in Figure 10. As before, these data represent two runs on the same specimen at different absolute crack lengths. In

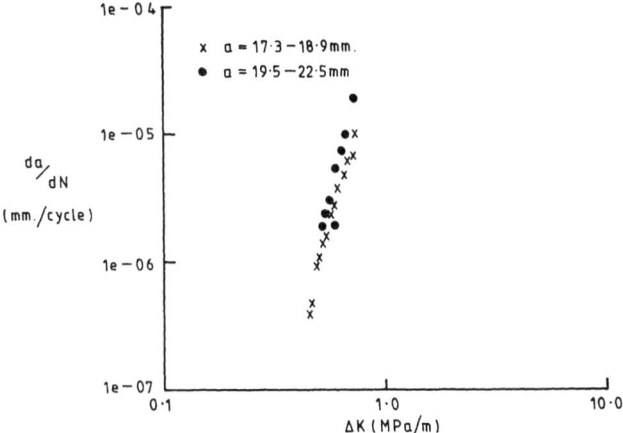

FIGURE 9. Fatigue behaviour of 0·5mm. bondline specimen

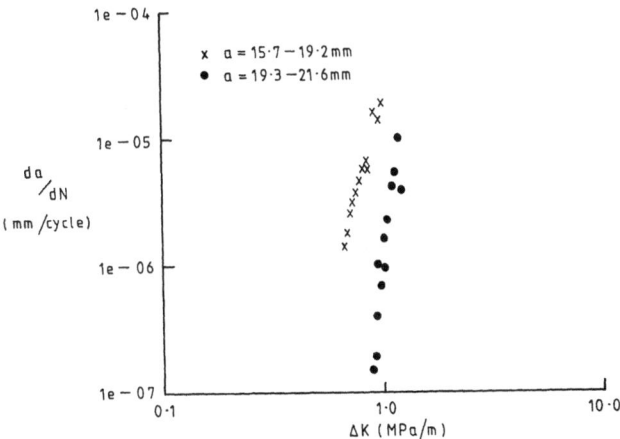

FIGURE 10. Fatigue behaviour of 2mm. bondline specimen

198

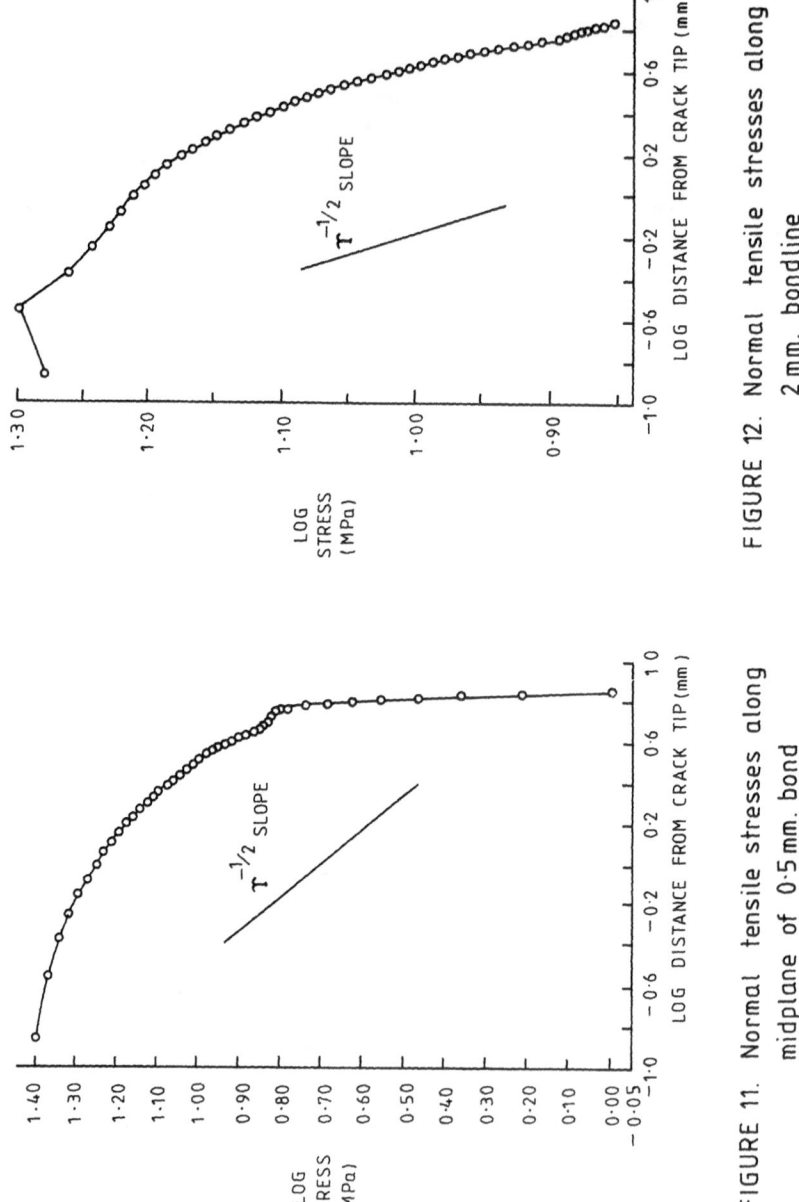

FIGURE 11. Normal tensile stresses along
midplane of 0·5 mm. bond

FIGURE 12. Normal tensile stresses along
2 mm. bondline

contrast to the 0.5 mm bond line case, there is little agreement between
the two evaluations. A single value of ΔK indicates that crack growth
rates may vary by an order of magnitude depending on the absolute crack
length in the specimen.

The use of the linear elastic stress intensity for characterising
crack tip stresses, and thereby failure, inherently assumes that the crack
tip stresses have an $r^{-\frac{1}{2}}$ singularity (11). Elastic finite element
analyses of the near crack tip stress fields have suggested that the
volume over which such a relationship exists may be limited (12). Plots
of the stress at the crack tip in a direction perpendicular to the crack
plane are shown for the 0.5 mm case in Figure 11 and the 2.0 mm case in
Figure 12. It is apparent that in the thin bond line case the stress
distribution can be approximated to an $r^{-\frac{1}{2}}$ singularity. For the thicker
bond line case, however, the region over which such a singularity exists
is limited. It is therefore not surprising that for the thin bond line
case crack growth rates may be correlated against ΔK whereas discrepancies
in behaviour appear when this is attempted in the thicker bond line
specimen.

CONCLUSIONS

1. The current analysis indicates that the cyclic strain energy release
    rate does not provide a unique parameter for defining fatigue crack
    propagation in bonded aluminium.

2. Finite element analyses have provided stress intensity calibration
    equations. Detailed examination of the near crack tip stresses
    indicates that for a 0.5 mm bond line thickness the stress
    singularity is dominated by an $r^{-\frac{1}{2}}$ dependence, in the 2.0 mm bond
    line thickness the stress singularity deviates from $r^{-\frac{1}{2}}$.

Acknowledgement
The authors would like to thank Alcan International Limited, Banbury
for permission to publish this work.

REFERENCES

1. A.J. Kinloch and S.J. Shaw, J. Adhesion (1981), 12, p.59.

2. H. Chai, (1986), ASTM STP 893, p.209.

3. K.M. Liechti, (1986), Structural Adhesives in Engineering, Proc. Inst. Mech. Eng., Bristol, July 1986, p.83.

4. P.D. Mangalgiri, W.S. Johnson and R.A. Everett, (1986), NASA Technical Memorandum 88992.

5. R.J. Bucci, W.G. Clark and P.C. Paris, (1972), ASTM STP 513, p.177.

6. ANSYS Users Manual, Swanson Analysis Systems Inc., Houston, PA, USA.

7. D.M. Tracey (1974), Nuclear Eng. and Design, $\underline{26}$, p.282.

8. ASTM Standard E399-83.

9. F. Erdogan and G. Gupta, (1971), Int. J. Solid Structures, $\underline{7}$, p.39.

10. P.D. Hilton and G.C. Sih (1972), Proc. 6th S.E. Conf. on Theoretical and Applied Mechanics, $\underline{6}$, p.949.

11. J.F. Knott, Fundamentals of Fracture Mechanics, Butterworths, London, 1973.

12. S.S. Wang, J.F. Mandell and F.J. McGarry, Int. J. Fract., $\underline{14}$, p.39, (1978).

13

Effects of the environment on bonded aluminium joints:

an examination by electron microscopy

J.A. Bishopp

Bonded Structures, Ciba-Geigy Plastics, Duxford, Cambridge, England

E.K. Sim, G.E.Thompson, G.C.Wood

Corrosion and Protection Centre, University of Manchester
Institute of Science & Technology, Manchester, England

INTRODUCTION

The use of scanning electron microscopy (SEM), to identify areas
of interest both on the substrate as well as on the adhesive
fracture surface, can be augmented by revealing the
adhesive/adherend interface via ultramicrotomy (1), with
subsequent observation by transmission electron microscopy (TEM).
This combination of techniques has already been established as a
valuable tool to characterise and gain a deeper insight into the
nature of substrates and interfaces within bonded joints (2, 3,
4, 5, 6).

The early data on one of the many aspects requiring more precise
characterisation, namely the effects of substrate pretreatment on
the resultant bond strength and its durability upon exposure to
"hot/wet" environments, are reported here.

THE ADHESIVE JOINT

In this work the adhesive joint has been formed by bonding Alclad
2024-T3 aluminium alloy substrates with an experimental,
toughened, supported, structural film adhesive curing at 120°C.

The aluminium pretreatments used are:

a)   Light" abrasion using Scotchbrite[R] or wire wool.

b)   Heavy" abrasion using an alumina grit blast.

c)   Potassium dichromate/sulphuric acid pickle to
     DTD 915b(ii) (7) [or in accordance with the Forest Products
     Laboratory process (8)]

d)   Potassium dichromate/sulphuric acid pickle, followed by chromic
      acid anodising (CAA) to DEF STAN 03-24/1 (9)

e)   Phosphoric acid anodising (PAA) to BAC 5555 (10)

In the latter stages of this work, the pretreated aluminium
adherends were also primed prior to bonding. The primer was an
experimental epoxy/phenolic system, heavily pigmented with
strontium chromate; full cure could be effected in 30 minutes
at 120°C.

ENVIRONMENTAL EXPOSURE

Within the limited time-scale that can be afforded to such an
analysis, an artificial environment has been used which
effectively accelerates the natural weathering effect. Thus,
floating-roller peel specimens (11) were exposed to a minimum 85%
relative humidity environment at 70°C for 30 days. Peel
strengths were determined every 10 days, but specimens for
characterisation work were only taken after 0 (Control) and 30
days exposure.

Mechanical Properties of Unprimed Joints

The peel profile for the exposed specimens is shown in Figure 1;
this reveals the differences in the peel strengths of the control
specimens and the reduction in peel properties after 30-days
exposure.

The degree to which surface pretreatment affects bond strength
and durability, are now discussed.

Controls: Figure 1 clearly shows that the peel values generated
fall into two groups; those where the substrates had been
abraded and those where chemical pretreatment had been used.

Examination of the adherend surface, the adhesive fracture
surface and the adhesive/adherend interface provides a possible
explanation for the varied behaviour. SEM analysis of abraded
substrates (Figure 2) shows that during pretreatment, although
the weak, air-formed film is removed, the aluminium cladding (5%
of total plate thickness) is partially cut open, producing a
relatively rough surface to which a significant amount of
aluminium detritus is loosely attached. SEM examination of the
fracture surfaces shows significant areas of apparent adhesion
failure to the peeling face, as well as evidence of air blisters
in the glueline (Figure 3). Analysis of transmission electron
micrographs of ultramicrotomed section permits one possible
hypothesis for the presence of such blisters to be postulated:
the surface, as prepared, is too rough. As the adhesive melts
and flows during its cure cycle, the air trapped between adhesive
and adherend should be displaced allowing the substrates to be
fully wetted. Figure 4 indicates that, because of the high
degree of roughness, not all the air has been displaced. This
has left air pockets, and consequently macroscopic areas of
either no contact or point-contact, between adhesive and
adherend.

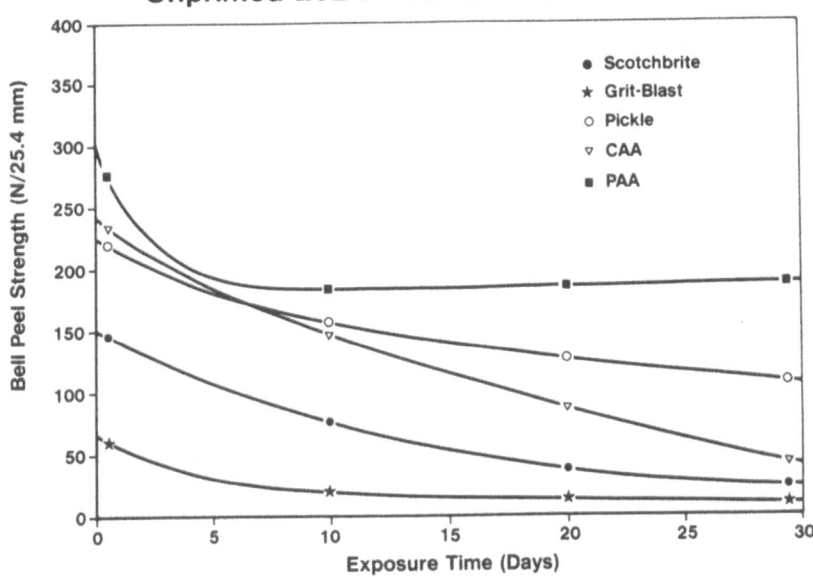

Unprimed 2024 – T3 ALCLAD Adherends

Legend:
- ● Scotchbrite
- ★ Grit-Blast
- ○ Pickle
- ▽ CAA
- ■ PAA

Figure 1:   Peel Strength versus Exposure Time at 70 Deg C and 85% R.H.
             for the various substrate surface pretreatments used.
             [Maximum scatter for any set of joints: +/- 15 N]

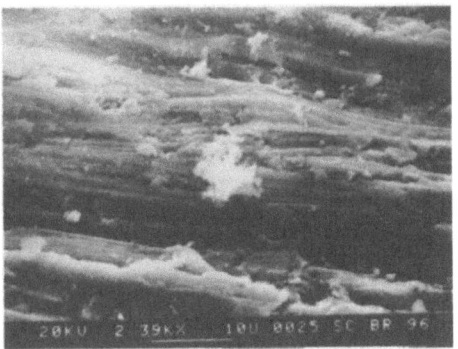

Figure 2:   Scanning Electron Micrograph of a Scotchbrite abraded
             substrate

204

Characterisation of the unexposed abraded bonded joint

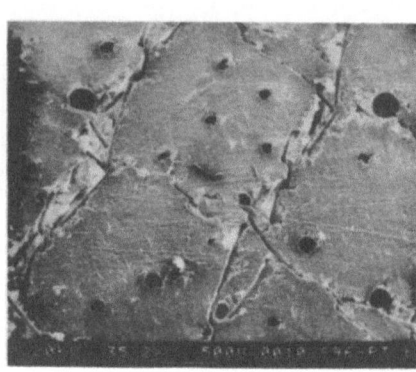

Figure 3:   SEM Micrograph of the
            adhesive fracture surface
            showing air-blisters in
            the glueline

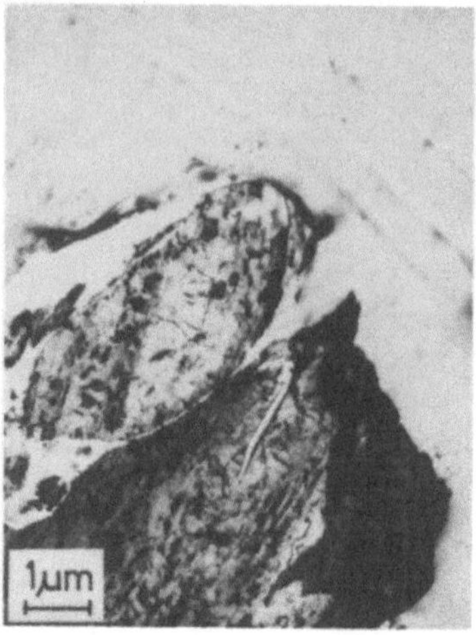

Figure 4:   TEM Micrograph of a
            section through the
            bonded joint showing
            poor wetting and
            surface detritus

Characterisation of the unexposed chemically pretreated bonded joint
----

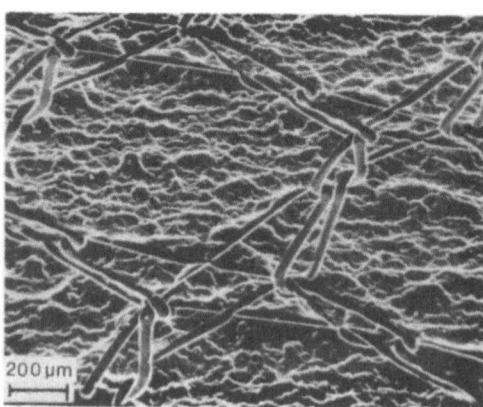

Figure 5:   Scanning Electron Micrograph of the adhesive fracture
            surface of a chrome/sulphuric pickled joint

Figure 4 clearly shows, also, the potential weakening effect on the unruptured joint by loosely bound surface detritus. In addition, a crack is also evident, running through the substrate immediately below the pretreated surface. One of the many possible explanations for this sub-surface fracture, that must be given serious consideration, is that stress-cracking in the adherend is induced by the method of pretreatment itself - in this case, grit-blasting.

The above suggests that no continuous, intimate interface exists between adhesive and adherend and hence failure should occur at low loads, as indeed is found. The so-called phenomenon of "mechanical keying" obviously plays little or no part in this instance.

SEM examination of the fractured, chemically pretreated joints shows, as expected from the peel strengths, significant areas of deep cohesive failure within the adhesive; Figure 5 is typical.

Use of ultramicrotomy to reveal the chemically pretreated adherend/adhesive interface shows, on examination by TEM, three very different morphologies:

i)   Potassium dichromate/sulphuric acid pickled surface (Figure 6): ostensibly finely-spaced, approximately 30 mm high whiskers which, from the evidence in the micrographs, appear to be well-penetrated by the adhesive. These whiskers, which are probably partially hydrated alumina contaminated by electrolyte species, are developed by transformation of the air-formed film during the immersion.

ii) Chromic acid anodised surface (Figures 7 and 8): a porous surface film is present which, in the work presented here, was generally 2 - 4 micrometres thick. The original surface roughness is not significantly enhanced by the anodising process and essentially mirrors the initial surface topography. Although the adhesive wets this surface well, there is no obvious deep penetration of the oxide pores by the adhesive.

iii) Phosphoric acid anodised surface (Figure 9): again, a porous surface film is present; generally about 0.5 - 1.0 micrometres thick. Here surface roughness is enhanced due to film material collapse during anodising (caused by progressive thinning of the cell material adjacent to the pore wall); this more open structure is also reported by W. Brockman et al (12). The degree of roughening, however, is some orders of magnitude less than for the abraded surfaces, which can be seen by the naked eye. The adhesive not only wets well, filling the revealed surface cavities, but careful examination of the enlarged micrograph also reveals evidence of pore penetration into the depths of the anodic film; this is confirmed by the work of others (13).

The use of Auger electron spectroscopy (AES) on the respective anodised samples shows that, whilst there is no evidence, within the detection limits, of any chromium associated with the CAA film, phosphorus is present throughout the section of the PAA film.

Characterisation of the unexposed chemically pretreated bonded joint
Transmission Electron Micrographs of sections through the joint
----

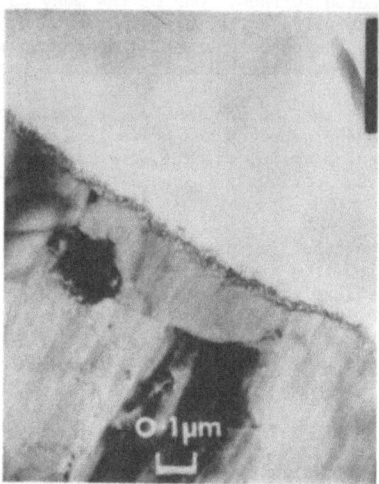

Figure 6:   Chromate/Sulphuric acid pickled to DTD 915b (ii)

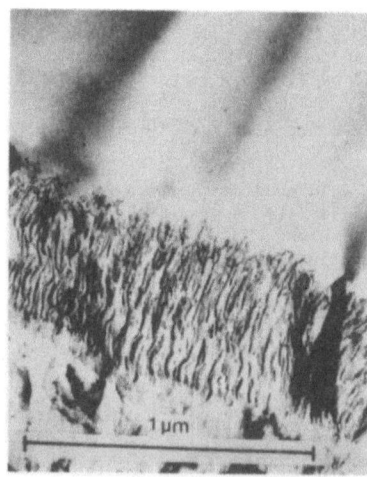

Figure 9:   Phosphoric acid anodised to BAC 5555

Characterisation of the unexposed chemically pretreated bonded joint
Transmission Electron Micrographs of sections through the joint
----

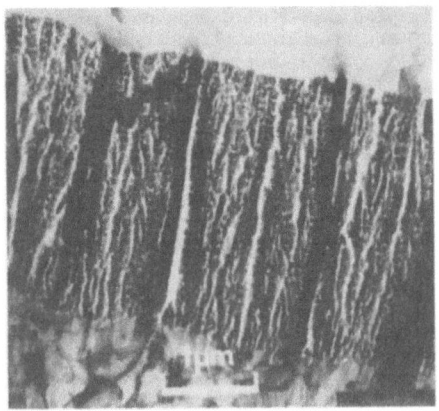

Figure 7:   Chromic acid anodised to DEF STAN 03-24/1

Figure 8:   Chromic acid anodised to DEF STAN 03-24/1

The interfacial examination has shown that, in all cases, the chemically pretreated adherend is wetted well by the adhesive; this, according to the theories of Zisman (14) should account for the relatively high peel strength levels on the substrates. Any increase in the effective surface area, by surface cavity formation in the case of the PAA adherends and by oxide film penetration in the case of pickled and PAA substrates, does not appear to affect significantly the average metal/metal bond strength for this adhesive. This would indicate that mechanical keying is not an important criterion in obtaining high peel loads. It could, however, explain, at least partially, the consistency of the peel levels for the PAA and pickled specimens as opposed to the slight variability observed on testing the CAA joints.

30-Day Exposure: The peel strength profile (Figure 1) indicates that the influence of substrate pretreatment on environmental exposure is much more varied than for the unexposed specimens.

Simple visual examination of the ruptured joints (Figure 10) reveals the areas in which significant environmental attack has apparently taken place. This is evident down the length of the joint, parallel and just adjacent to the cut edges (Figure 10C) and occasionally, dependent on pretreatment, extending towards the centre or even across the joint (Figure 10D). This latter tendency is particularly prevalent when abraded substrates are used.

It should be noted that the initial edge effect is likely to stem from damage to the interface, inflicted when sawing the joint prior to exposure. Strong support for this is given by observations of the uncut joint ends, protected by adhesive squeeze-out, which rarely show such attack.

SEM examination of specimens taken from the areas of apparent bond deterioration suggests, irrespective of surface pretreatment, failure at the adhesive/adherend interface; significant areas of apparent adhesion failure to the peeling substrate being evident.

TEM examination of the ultramicrotomed sections confirms the presence of extensive adhesion failure to the peeling face (Figures 11 and 12 are typical of all pretreatments); small areas of cohesive failure in the adhesive can be seen on the chemically pretreated substrates, whilst the abraded adherends appear to exhibit almost total adhesion failure.

The so-called pseudoboehmite morphology was clearly seen on these exposed aluminium structures, remote from the unfractured area.

However, specimens taken in advance of the crack tip (Figures 13 - 16) i.e. in the unruptured area of the joint, show no evidence of pseudoboehmite formation at the interface. Hence, in this case, this well-characterised form of hydrated alumina is formed mainly post-rupture and is not the cause of failure per se.

209

Characterisation of the environmentally exposed bonded joint

Figure 10:   Visual examination of ruptured peel joints:-

A - Unexposed control

C - 30 Days exposure: edge
    attack

B - 30 Days exposure no obvious
    attack

D - 30 Days exposure: extensive
    attack

Transmission electron micrographs of sections through
the peeling face of ruptured peel specimens

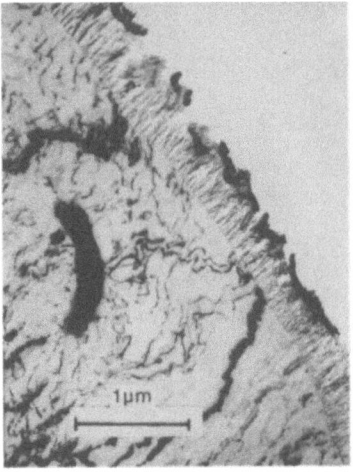

Figure 11:   Scotchbrite abraded
             substrate

Figure 12:   Phosphoric acid
             anodised substrate

Characterisation of the environmentally exposed bonded joint
Transmission Electron Micrographs of sections through the
unruptured bonded joint after 30 days environmental exposure

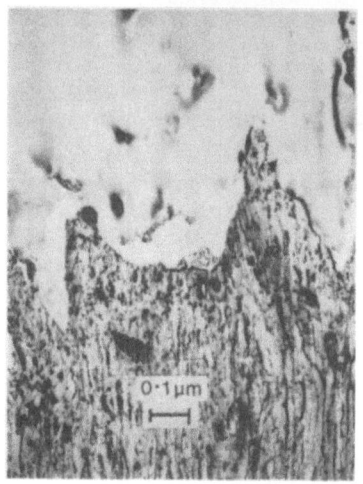

Figure 13:　　Alumina grit-blasted
　　　　　　　　substrate

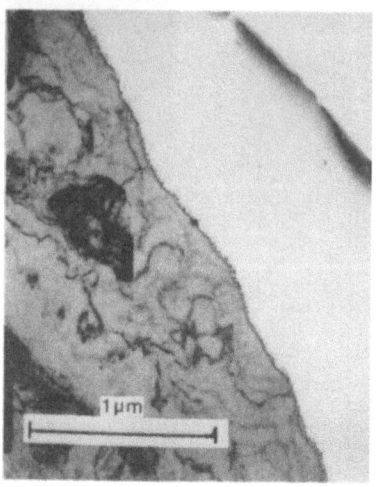

Figure 14:　　Chromate/Sulphuric
　　　　　　　　acid pickled
　　　　　　　　substrate

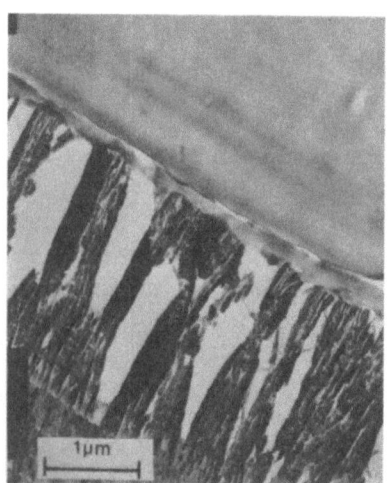

Figure 15:　　Chromic acid
　　　　　　　　anodised substrate

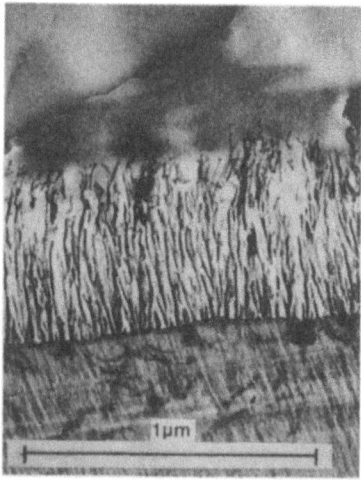

Figure 16:　　Phosphoric acid
　　　　　　　　anodised substrate

Figures 13 - 16 show little evidence of significant degradation at
the pre-rupture interface, or in the adherend immediately under the
interface. It can, therefore, be postulated that the reduction in
peel strength is due to an ingress of water either weakening or
causing a debonding effect at or immediately adjacent to the
adhesive/adherend interface, or a transformation of the surface
film so subtle as not to be identifiable by the analytical
techniques being used.

There are three possible paths by which water can diffuse through
the joint : i) through the adhesive itself, assisted by cracks
and voids in the organic matrix or by wicking along the carrier/
adhesive interface, ii) through the grown or air-formed film on
the adherend surface or iii) along the adhesive/adherend
interface. In all cases water ingress would have to start at the
exposed joint edges and thus, in view of the length of the
diffusion path to the interior of the joint, any immediate attack
must start at or closely adjacent to the edge of the bonded area.

As can be seen in Figure 10D, when bonded specimens are produced
using abraded substrates, this initial edge attack is frequently
augmented by rapid ingress into the joint. The areas of
non-contact between adhesive and adherend, caused by the
over-rough surface and the loosely-attached detritus of
pretreatment (Figure 4), offer a more direct route through the
joint and, hence, water should be able both to enter and pass
into the interior of the bonded area with relative ease.
This would effectively short-circuit the potentially much longer
path through the adhesive and, obviously, allow water to be taken
up at a much higher rate than along the interface of a well-wetted
adherend. The already poor adhesion, due to the nature of the
surface and the unsuitability of the air-formed film which is
produced rapidly after pretreatment, is thus further reduced both
by possible over plasticisation of the adhesive matrix by water
and feasibly by the disruption of the physico-chemical bonds
Van de Waal's forces, etc.) across the interface.

When chemically pretreated adherends are used an even more subtle
than previous mechanism of attack must operate. The TEM
micrographs and control peel strengths confirm that these
surfaces are wetted well by the adhesive. There is, therefore,
no such easy route for water to reach the interior of the joint
and, indeed, any immediate edge effect is generally not rapidly
followed by environmental attack across the bonded area.

Clearly, water eventually penetrates the bonded area since
significant reductions in peel strength are observed and it could
be argued that, because of the good wetting of the adherend by
the adhesive, the dominant diffusion path is probably through the
adhesive matrix.

Deep cohesive failure, in the adhesive, is rarely evident in
ruptured joints which have previously undergone environmental
exposure. This would indicate that the water has caused
relatively little damage to the bulk of the adhesive matrix
compared with the more significant processes proceeding at or
close to the interface.

In the following, although only PAA and CAA pretreated adherends
are specifically considered, it is reasonable to treat the
pickled surfaces as behaving somewhat similarly to those anodised
in phosphoric acid. However, for pickled aluminium the "pore"
depth is only about 3-4% of that of PAA substrates and thus, in
this case, the film/metal interface is significantly closer to
the outer film surface and hence, could play a more significant
role in the mechanism of bond degradation.

The water ingress, once at the interface, may wet the adherend
surface and porous structure comprising the anodic film
morphology better than the adhesive matrix already in place; the
extent of pore wetting will be dependent on the degree of
adhesive penetration. In such cases, this wetting of the alumina
material may be sufficient, per se, for bond deterioration by
significantly reducing the Van der Waal forces etc., across the
interface. If not, then a subtle transformation of the adjacent
alumina film to AlOOH and $Al(OH)_3$, at a rate dependent on the
exposure conditions, must be considered.

Speculating further, in situations where the pore volume is not
penetrated substantially by the adhesive, wetting of the pore
surface and possible build-up of water and other potentially
damaging species, within the pore, (either direct from the
environment or dissolved out of the cured adhesive matrix, as
shown for example by Brockmann et al (12), can also be
contemplated; a ready transformation of the anodic alumina to
hydrated material would be anticipated. Transformation is
expected to proceed by penetration of the cell material
surrounding the pore, developing a disagreggated alumina zone
behind which dissolution and reprecipitation occur. This
disruption of the alumina cell, with precipitation of relatively
voluminous hydrated material, could contribute significantly to
the undermining of the anodic film surface and hence lead to
bond deterioration.

Further, should moisture gain access to the pore volume then
direct passage to the metal, via flaws in the substrate, is a
distinct possibility. Once in contact with the metal, water,
assisted by damaging species (e.g. chloride ions, ammoniacal
materials), can enhance corrosion of the substrate, undermining
the alumina film. Considering the situation when the adhesive
substantially penetrates the porous morphology, any undermining
of the alumina (through hydration proceeding into the cell walls)
and corrosion of the metal are likely to be significantly
delayed.

However, in the present study, where durability has been
assessed by exposure to relatively high humidity, examination of
the adhesive fracture surface shows no strong evidence for failure
through the alumina film and hence, crack propagation should
proceed along, or close to, the interface, with intermittent
diversions into the immediately adjacent bulk of the adhesive
matrix.

Further, hydrated alumina is not readily observed attached to the adhesive fracture surface. This indicates that the degree of any surface transformation does not have to be extensive in order to cause potentially severe bond degradation.

In the work described here, the adhesive, metal adherends, cure cycle and operators have, essentially, been kept constant. Thus, the differences revealed in the resistance to environmental attack by the bonded specimens, prepared using chemical pretreatments, must be due to the pretreatments themselves.

Before these differences can be fully explained, however, the moisture diffusion path, the factors affecting the rate of diffusion and the alumina/moisture reaction must be more fully understood. Some pointers do already exist:

i)   As opposed to phosphoric acid anodised surfaces, those produced in chromic acid, although possessing a porous morphology, reveal a macroscopic planar surface in intimate contact with the adhesive but with little pore penetration (Figures 7 and 8).

The outer regions of the PAA film, however, comprise both a coarse and a fine cavity-like structure, due to film collapse (Figure 9); these areas are readily wetted by the adhesive, which also occupies a significant fracture of the internal film volume. Thus, for such anodic films, the moisture diffusion paths are considerably extended, when compared with those for CAA film.

The above argument will also generally apply to the joints where pickled substrates have been used; the adhesive wetting and, to a large extent, penetrating the whisker-like film growth and hence extending the interfacial area and eliminating any real planar interface.

Considering the PAA adherends further, the presence of bound, absorbed and adsorbed phosphates could well mask potential hydration sites, slowing down the damaging transformation to hydrated alumina. However, even in the presence of these phosphates, if hydration does eventually proceed in the outer regions of the anodic film then the transformed areas will be almost fully enveloped by adhesive. In other words no distinct, weak interphase can develop parallel to the aluminium substrate surface.

In addition, any later corrosion processes in the metal itself are also likely to be delayed because of the penetration of adhesive into the pores.

ii) Although it has been stated that little indication of interfacial damage had been found by TEM/Ultramicrotomy examination, there was a specific area where some evidence of hydration existed; on pickled and abraded substrates at the adherend/adhesive interface, in the immediate vicinity of second-phase particles in the aluminium cladding. If such inclusions are resident at or very close to the surface (i.e. at locations where they are readily revealed by pretreatment) then they may exert an influence on the performance of the joint under environmental exposure. Such an effect should, therefore, be limited to joints where the adherends have either been mechanically abraded or pickled. This has, indeed, been found; Figures 17 - 19 show the enhanced local degradation of surface films in the presence of those intermetallic inclusions.

If water, containing any dissolved electrolytes species, reaches the interface, then, as the intermetallics are essentially of the $FeAl_3$ type the possibility of galvanic corrosion, with the adjacent aluminium matrix serving as the anode, must be considered. However, a further possible cause for this interfacial degradation could be weak (in the bonding sense), readily hydrolyseable films developing above the inclusion itself; as the interface recedes there is the possibility of new inter-metallics being revealed and an easier environmental pathway being opened up to them.

Although requiring further clarification, this effect must be put into perspective. The cladding contains a _maximum_ 0.7 wt % iron and not all the inclusions will be sufficiently close to the surface to exert an influence. However, any local surface degradation caused as a result of the presence of intermetallics may facilitate the passage of water along the interface, which would then, indeed, have a significant effect on joint durability.

Mechanical Properties of Primed Joints

Work has commenced on this examination and is already showing results of considerable interest. With abraded adherends, it appears that the primer wets and fills the smaller troughs and craters of the convoluted surface and assists in "fixing" some of the loosely-bound surface detritus (Figure 20).

For the chemically pretreated substrates, the question of primer penetration of the film surface through the pores is of major importance. It is thought that there is little difficulty in the case of pickled adherends. Figure 21 shows that some component, at least, of the primer does penetrate the PAA film. This has been confirmed using an organotitanate labelled primer (15); titanium was detected largely throughout the porous film morphology. This same work has also shown that some component of the primer does partially penetrate the CAA film; titanium was detected down to about one-third of the pore depth.

The peel strength profile (Figure 22) shows, once again, the marked difference between specimens prepared with abraded substrates and those using chemical pretreatments.

For the abraded adherends, some improvement in durability is to be expected through the presence of the primer and its pigment particles. The latter produce soluble chromate species which, upon migration to the air-formed film surface will hinder hydration. Further, if corrosion in the vicinity of

Characterisation of the environmentally exposed bonded joint
The effect of intermetallic inclusions on environmental attack
Transmission Electron Micrographs of sections through the
unruptured bonded joint after 30 days environmental exposure

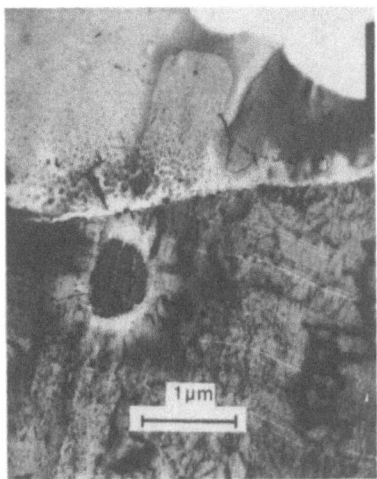

Figure 17:   Alumina grit-blasted   Figure 18:   Chromate/Sulphuric
              substrate   acid pickled
                                                       substrate

Figure 19:   Chromate/Sulphuric acid pickled substrate

Characterisation of the unexposed bonded joint having primed substrates
Transmission Electron Micrographs of sections through unruptured joints

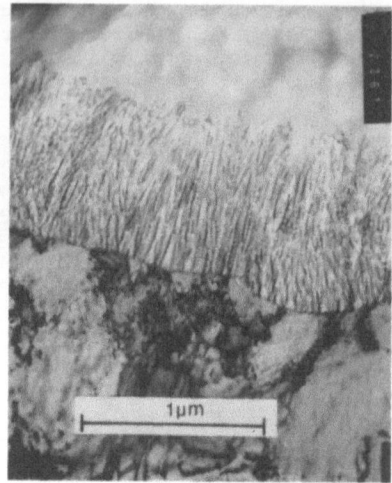

Figure 20: Alumina grit-blasted substrate

Figure 21: Phosphoric acid anodised substrate

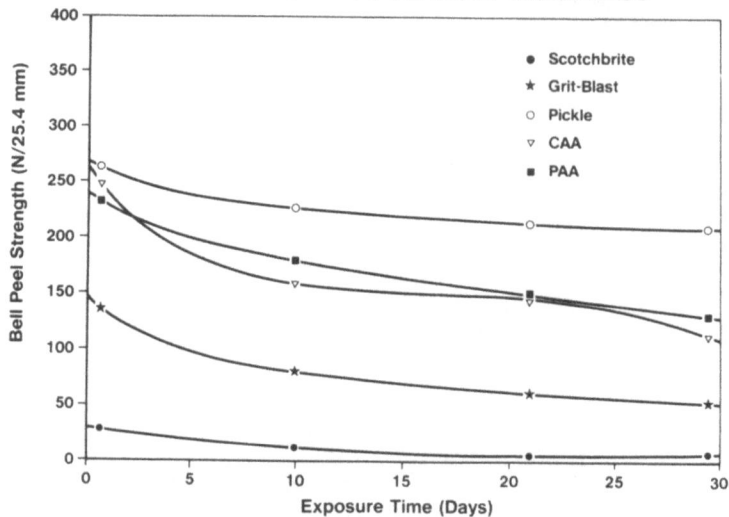

Figure 22: Peel Strength versus Exposure Time at 70 Deg C and 85% R.H. for the various primed substrate surface pretreatments used [Maximum scatter for any set of joints: +/- 10 N]

intermetallic and second-phase material is a significant factor in bond degradation, then chromate species are highly effective inhibitors of aluminium corrosion. However, all these factors may be largely outweighed by the still restricted wetting of adherends with complex geometries; this would appear to be confirmed by the peel profile which shows relatively rapid bond deterioration.

Similarly, for the chemically pretreated substrates, an improvement in durability is anticipated. Generally, the primer or its components penetrate the pore volume, thereby enhancing the area of contact. Consequently, hydration by water - contaminated or otherwise - contained within the porous morphology, is hindered for the reasons outlined previously (for adhesive penetration of the oxide film morphology); released chromate species should also limit hydration to voluminous film material.

Finally, if good wetting of the adherend surface coupled with adhesive/primer penetration of the oxide film morphology is the key to limiting the environmental attack on the bonded joint (by increasing the diffusion time of water to the interface), then it would be expected that:

i)   There should be little difference in durability between primed or unprimed joints using PAA adherends.

ii)   Priming CAA substrates should improve joint durability towards that of the PAA system.

iii) If the primer can seal off the intermetallics from ready access to moisture ingress, then an improvement in durability is to be expected on priming pickled adherends.

The early results are, essentially, in support of these statements.

CONCLUSIONS

The following conclusions can be drawn from the work reported here:

1)   For good wetting of the adherend, and hence good adhesive strength levels, chemically pretreated substrates appear essential. The surface convolutions and loosely-bound detritus, produced by mechanical abrasion, lead to low levels of intimate contact, weak interfacial layers and stress-cracking in the adherend; all contributing to poor bond strength.

2)   The needle-like oxide morphology formed on pickled substrates and the rough, porous anodic oxide layer grown in phosphoric acid appear to be well penetrated by the adhesive; although porous, the CAA film does not appear to be significantly penetrated by adhesive.

218

3) A significant level of incorporated and adsorbed phosphate is found throughout the PAA film. Within the detection limits of AES, no parallel is evident for the CAA or pickled substrates, although anion (chromate) adsorption on exposed surfaces is expected.

4) On unprimed substrates, the effect of pretreatment on the resistance to environmental attack can be rated, in ascending order; grit-blasting and "light" abrasion <chromic acid anodising <pickling<phosphoric acid anodising.

5) The morphologically characterised pseudoboehmite form of hydrated alumina appears to be formed only _after_ rupture and hence is not thought to be the cause of environmental failure per se.

6) Environmental attack appears either to disrupt the physico-chemical bonds across the interface or very subtly to undermine the interface itself, rather than weakening the adhesive matrix.

7) Penetration of the adhesive into the depths of the oxide film is important in that it effectively increases the length of the interface at the same time as eliminating a true planar boundary between adhesive and adherend. Thus, for pickled and PAA adherends, once water reaches the interface, there is no continuous passage for moisture between adhesive and adherend - effectively hindering any possible bond disruption, and should transformation of the alumina film, to its hydrated form, take place then any area attacked will be isolated by the surrounding adhesive. Both will limit the development of a continuous, weak interphase.

For CAA adherends, a planar interface exists with little adhesive penetration into the pores; any environmental attack can, therefore, proceed more rapidly both causing bond disruption and having the potential for the undermining of the anodic film surface.

8) If water can reach the surface film/metal interface, through flaws or by means of intermetallic sites, then corrosion is feasible, unless limited by the presence of primer. The result again would be an undermining of the pre-developed "oxide" films. This would be particularly relevant where the developed film is relatively thin - i.e. on abraded or pickled adherends.

ACKNOWLEDGEMENTS

The authors gratefully acknowledge the help of the following:

Dr. P. Wilford, RAE, Farnborough for releasing early data of RAE/UMIST studies on pretreated aluminium adherends.

BAe, Weybridge and BAe, Hatfield, for providing phosphoric acid anodising facilities.

Mrs. S.C.Harrow and Miss J.A.Underwood for preparing all the bonded specimens and carrying out the environmental testing.

REFERENCES

1. Furneaux, R.C., Thompson, G.E. and Wood, G.C., Corrosion Science, 18, p 853, 1978

2. Smith, T.V., MSc Dissertation, University of Manchester Institute of Science and Technology, 1983

3. Bishopp, J.A., International Journal of Adhesion and Adhesives, 4, p 153, 1984

4. Bishopp, J.A., Smith, T.V., Thompson, G.E. and Wood, G.C., Proceedings of the 169th Meeting of the Electrochemical Society Inc., Boston, May 1986

5. Allen, K.W., ed., Adhesion 12, Elsevier Applied Science Publishers, Barking, England, p 248, 1987

6. Bishopp. J.A., Sim, E.K., Thompson, G.E., and Wood, G.C., Proceedings of the Plastics and Rubber Industries Conference - Adhesion '87, York, September 1987

7. Aircraft Process Specification, DTD 915B, Ministry of Supply, 1956

8. Eickner, H.W. and Schowalter, W.E., Report 1813, Forest Products Laboratory, Madison, Wisconsin, May 1950

9. Defence Specification, DEF STAN 03-24/1, "Chromic Acid Anodizing of Aluminium and Aluminium Alloys", Ministry of Defence, 1984

10. Boeing Process Specification, BAC 5555, "Phosphoric Acid Anodising of Aluminium for Structural Bonding", 1974

11. A.E.C.M.A. Standard pr EN 2243-2, August 1980, (Draft - Issue 1)

12. Brockmann,W., Hennemann, O.-D., Kollek, H., and Matz, C., Adhesion and Adhesives 6, No. 3, p 115, 1986

13. Skeldon, M., Thompson, G.E., Wilford, P. and Wood, G.C., "Adhesive Bonding of Aluminium Alloys", Final Report, 2044/0130 XR/MAT, 1987

14. Zisman, W.A., Industrial & Engineering Chemistry, 55, p 19, 1963

15 Private communication with Dr. P. Wilford, RAE, Farnborough 1987

APPENDIX

Composition of ALCLAD 2024- 3

	Percentage Composition	
	Core	Cladding
Copper	3.8 - 4.9	0.1
Magnesium	1.2 - 1.8	-
Manganese	0.3 - 0.9	0.05
Iron	0 - 0.5	)
Silicon	0 - 0.5	) 0.7
Chromium	0 - 0.1	-
Zinc	0 - 0.25	0.1
Others (individual)	0 - 0.05	0.05
Others (total)	0 - 0.15	0.15
Aluminium	Rest	Rest

14

THE STUDY OF STRAIN FIELDS IN ADHESIVE JOINTS BY LASER MOIRE INTERFEROMETRY

ROGER DAVIDSON
Materials Development Division,
Harwell Laboratory, Abingdon,
Oxfordshire, OX11 ORA, U.K.

INTRODUCTION

Until recent developments in experimental mechanics it has not been possible to measure the detailed deformation fields in stressed structural adhesive joints. The problem is made difficult because bond thicknesses are thin, typically less than 0.5 mm. In order to obtain the shear characteristics of adhesives, much effort has been made in developing special extensometry to measure the relative displacements across joints [1,2]. These are point deformations and give no information as to how the shear deformations vary along or through the adhesive bond. Also it is known from theoretical analyses that even when an adhesive appears to be subjected only to shear displacements, as in the thick adherend and shear test (TAST), there are always associated peel deformations which also vary along the bond. The application of laser moiré interferomety as refined by Post [3] offers the only known method of measuring on a very fine scale such full field displacement distributions in stressed adhesive joints. The sensitivity of typically 0.42 μm per fringe order allows such measurements to be used to:

- verify and refine theoretical models

- gain an improved understanding of the mechanisms which control failure

- quantitatively observe effects which have proved very difficult or
  even impossible to model accurately.

This paper will illustrate the value of LMI by consideration of the results
obtained in a thick adherend shear joint geometry produced by bonding steel
adherends with a structural epoxy adhesive. The overlap length has been
deliberately made longer than usual in order to give enhanced strain
contributions near the ends. The results will be compared to theoretical
predictions using a continuum mechanics approach, BISEPS-LOCO, developed at
Harwell.

PRINCIPLE OF MOIRÉ INTERFEROMETRY

Moiré techniques rely on a parallel grid, carried on the surface of a
specimen, which deforms as the state of strain within the specimen is
changed through for example mechanical stress or thermal effects.
Interrogation of the specimen grid through an underformed aligned reference
grid produces contour maps of equal displacement. These are moiré
fringes [4]. Traditional geometrical moiré is limited by the coarseness of
the finest grids achievable to a maximum of 40 lines per mm. Post [3,5,6]
first succeeded in demonstrating a usable technique up to two orders of
magnitude more sensitive. This is done by the utilization of two beam
interference of coherent laser light to produce both specimen gratings of
up to 2000 lines per mm and also virtual reference gratings of twice the
specimen grating frequency. The specimen grating is produced by exposing
high resolution holographic plates to the interfering laser light. By
rotating the plates through 90° between exposures a cross grating pattern
is imaged. On developing and bleaching, the relative shrinkage of the
exposed and unexposed parts of the plates produces a corrugated phase type
two dimensional cross diffraction grating. In the work described here the
gratings used had 1200 lines per mm.

Specimen gratings are replicated onto the sample to be tested by first
vacuum evaporating aluminium of thickness 0.025 mm onto the plate surface.
This aluminium coating follows the contours of the grating and is
transferred to the specimen via a room temperature curing, non-shrinkable
photoelastic epoxy adhesive. A small amount of the adhesive is placed
onto the prepared specimen surface and the aluminium coated grating,

aligned with respect to the specimen loading axis, is pressed against it to spread the epoxide into a thin film. Upon curing the mould is easily prized off to leave the replicated aluminium layer adhering to the specimen surface.

OPTICAL SYSTEM AND LOADING FRAME

The success and flexibility of the laser moiré technique relies upon the production of a virtual reference grating being generated in space which interacts with the specimen grating to produce deformation fringes which can be photographed and analysed. The experimental technique is illustrated in Figure 1. One coherent beam reaches the specimen directly at $-\alpha$ while the other is reflected by a plane mirror, orthogonal to the specimen, to reach the specimen at $+\alpha$. The two intersecting diffracted beams emerge approximately normal to the specimen surface and form alternate vertical lines of constructive and destructive interference on the specimen surface. The angle to satisfy these conditions is given by

$$\alpha = \sin^{-1}\left(\frac{f\lambda}{2}\right)$$

This corresponds to an angle of $49.4°$ for He–Ne laser light of wavelength 632.8 mm and a virtual grid frequency, f, of 2400 lines per mm.

Light from the two diffracted beams is collected by a large diameter objective lens focussed on the specimen surface and imaged on a camera back to record the moiré fringes. With the reference grating lines perpendicular to the x-axis the pattern is a contour map of displacements governed by the relationship

$$u = \frac{1}{f} Nx$$

where u is the x component of displacement at any point and Nx is the fringe order at that point in the fringe pattern.

In order to obtain the v displacements two mirrors orientated at $\pm45°$ to the vertical axis and orthogonal to the specimen surface are used to produce a horizontal vertical grating. This is illustrated in Figure 2. By using a sufficiently expanded laser beam of up to 150 mm the three

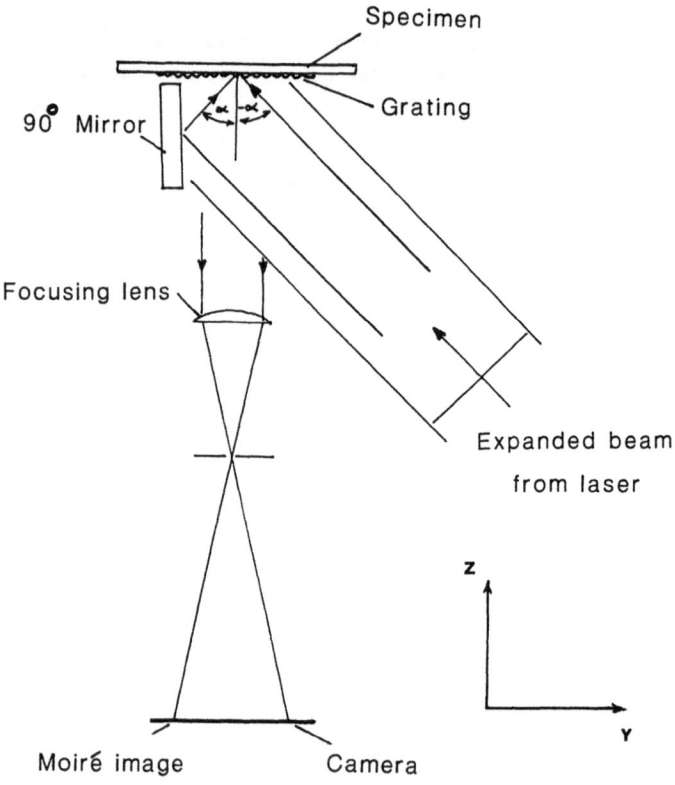

Figure 1   Optical   arrangement for laser moiré interferometry.

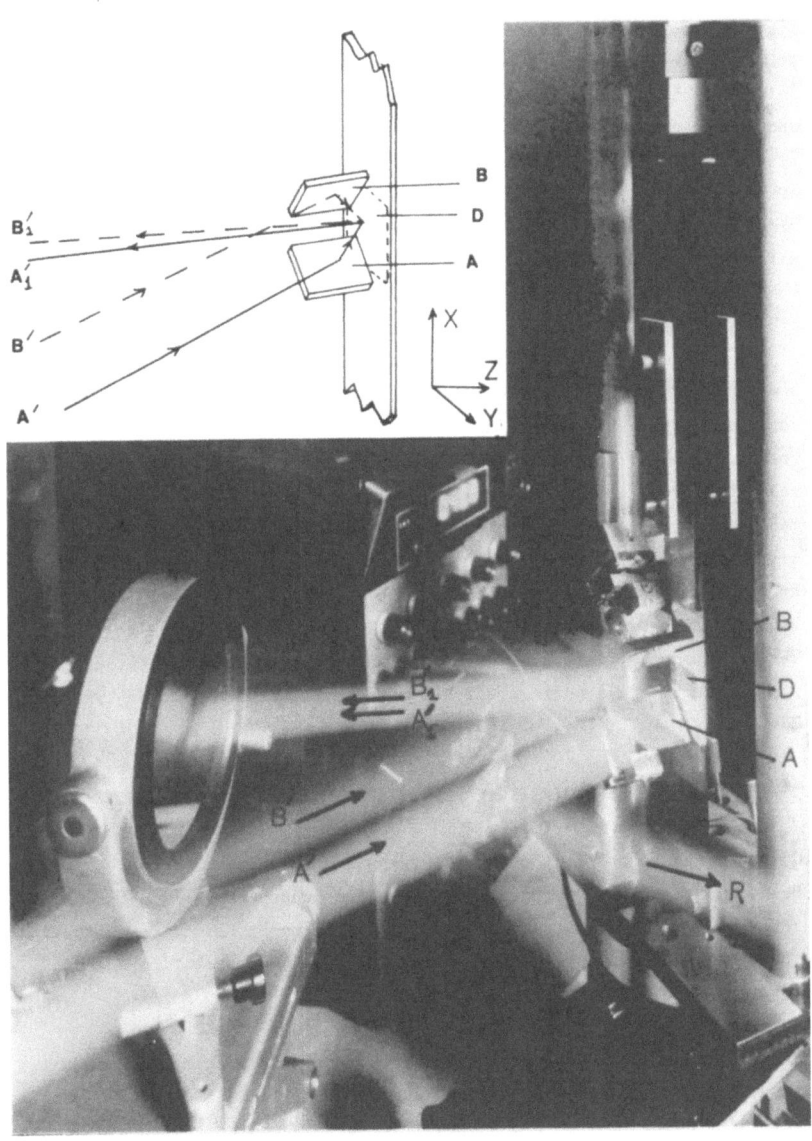

Figure 2   Laser moire interferometry apparatus, showing

the production of u fringes.

mirrors and the specimen can be simultaneously illuminated such that a stressed specimen area of several square centimeters can be studies. Selective illumination of the 90° mirror and the 45° mirrors allows the u and v deformation fringes to be obtained independently. The three mirrors used to split the laser beam are mounted into small goniometer stages attached firmly onto the loading frame. The vertical positions of the mirror holders are adjustable as are the angular positions of the mirrors, allowing the accurate positioning required to produce the aligned virtual reference grids. The laser moiré interferometry apparatus showing the production of u displacement fringes is shown in Figure 2.

A specially designed loading frame has been constructed to be used in conjunction with the optics described above. It is mounted on a tilt and rotation stage with the load applied through a worm drive mechanism and reacted by two 25 mm diameter aluminium alloy columns. The applied load is measured by a compact 10 kN load cell. The goniometer stages holding the mirrors are mounted on a thin walled unstressed aluminium cylinder held around one of the loading columns. The system is relatively insensitive to vibration and distinct fringes are obtained without the need for vibration resistant optical tables. The system has been successfully set up on standard triangular optical benches mounted on a sturdy wooden laboratory bench.

THE USE OF LASER MOIRÉ INTERFEROMETRY IN THE STUDY OF ADHESIVE JOINTS

Computer models to predict the stiffness and strength of adhesive joints, based on a continuum mechanics analysis and incorporating non-linear adhesive effects are under continued development at Harwell. Validation of these models requires a global map of the strain distribution in both adherends and adhesive layer.

In the present work laser moiré interferometry has been applied to the study of shear and peel strains in a structural epoxy adhesive. The TAST-like specimen used had 5.5 mm thick steel adherends with a distance between the cut outs of 19.4 mm and a bond thickness of 0.533 mm. The specimen illustrated in Figure 3 was set up in the loading frame and the mirrors adjusted to give approximately null fields in the unloaded specimen. The u and v deformation fringes were photographically recorded

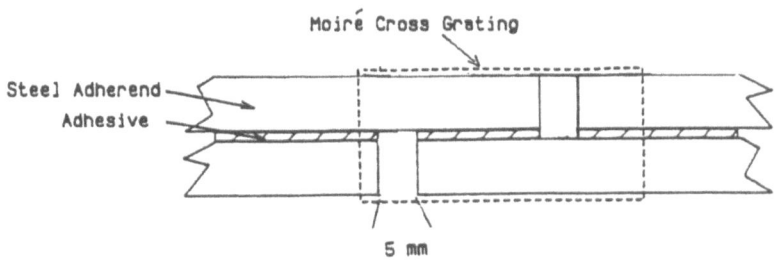

Moiré Cross Grating

Steel Adherend

Adhesive

5 mm

Figure 3   Schematic of TAST Specimen

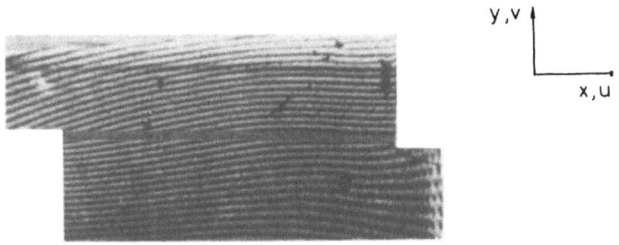

y,v

x,u

Figure 4   U-Moiré displacement fringes for TAST
                 Specimen loaded to 500 N (2.1 MPa)

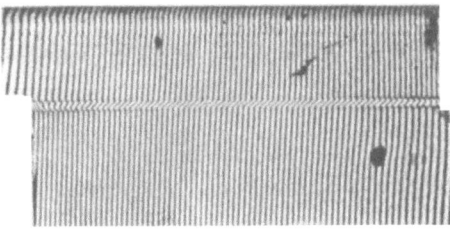

Figure 5   V-Moiré displacement fringes for TAST
                 specimen loaded to 1000 N (4.2 MPa)

in steps of 500 N to obtain both the elastic response and the initiation
and growth of damage within the bond.

RESULTS

The u displacement fringes in the steel adherends and in the adhesive for a
load of 500 N are shown in Figure 4. Fringes parallel to the adhesive bond
length are clearly observed with the fringe count across the adhesive
thickness increasing towards the cutouts. Transverse deformation fringes
along the bond length of the specimen at 1000 N are shown in Figure 5. The
fringe count across the thickness of the adhesive is small and negative at
the central regions of the joint becoming larger and positive close to the
cutouts.

U-displacement fringes at the central region of the joint for increasing
load are shown in Figure 6. The fringe counts at the centre of the joint,
corrected for the small count measured at zero load, are indicated by the
values of ΔN_c shown in Figure 6. The transverse displacement fringes from
the joint centre to one of the cutouts are shown for increasing loads in
Figure 7. Figure 7a shows the fringe distributions at 3000 N measured from
an approximately null field at zero load. In order to observe clearly the
effects at higher loads one of the 45° mirrors was adjusted slightly to
introduce a carrier pattern in order to coarsen the moiré fringes as shown
in Figure 7b and 7c.

DISCUSSION

Moiré fringes are contour lines of equal u and v displacements. To obtain
strain information it is necessary to apply the Langrangean large
deformation relationship given by:

$$\varepsilon_x = \frac{\partial u}{\partial x} + \frac{1}{2}((\frac{\partial u}{\partial x})^2 + (\frac{\partial v}{\partial x})^2)$$

$$\varepsilon_y = \frac{\partial v}{\partial y} + \frac{1}{2}((\frac{\partial u}{\partial y})^2 + (\frac{\partial v}{\partial y})^2)$$

$$\gamma_{xy} = \arcsin (\frac{\frac{\partial u}{\partial y} + \frac{\partial v}{\partial x}}{(1 + \varepsilon_x)(1 + \varepsilon_y)})$$

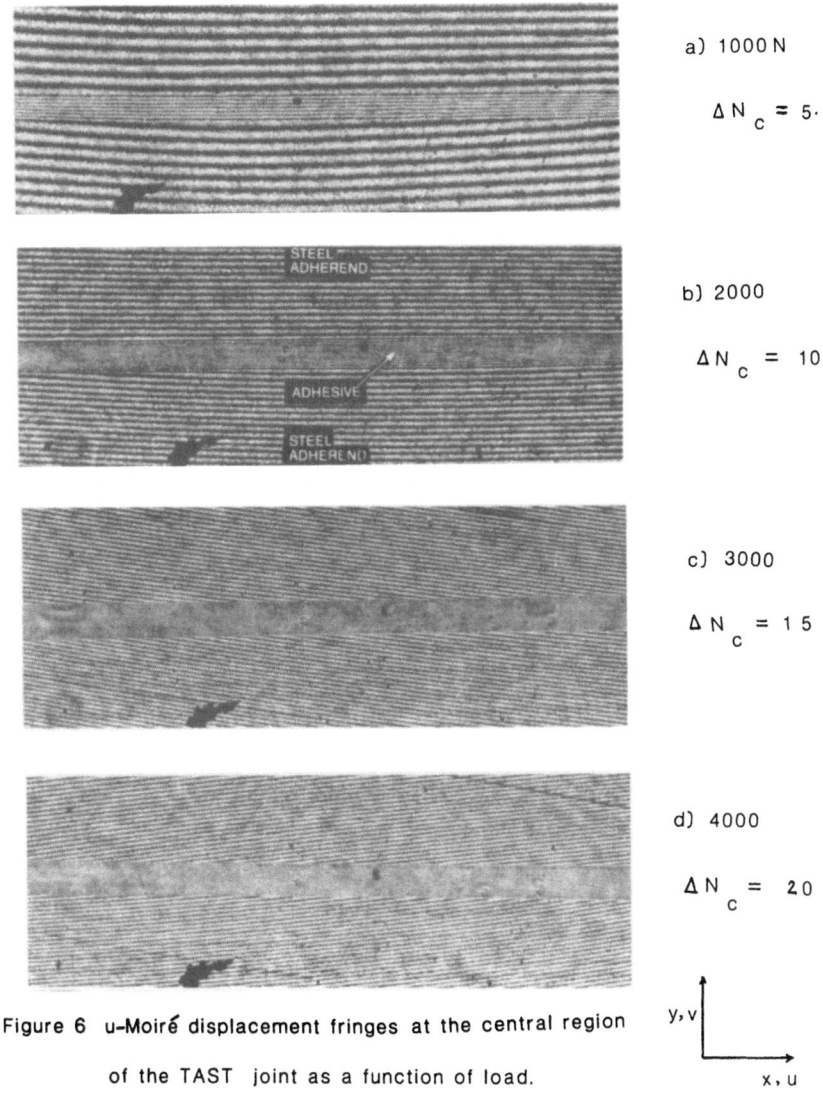

a) 1000 N

$\Delta N_c = 5.$

b) 2000

$\Delta N_c = 10$

c) 3000

$\Delta N_c = 15$

d) 4000

$\Delta N_c = 20$

Figure 6 u-Moiré displacement fringes at the central region
of the TAST joint as a function of load.

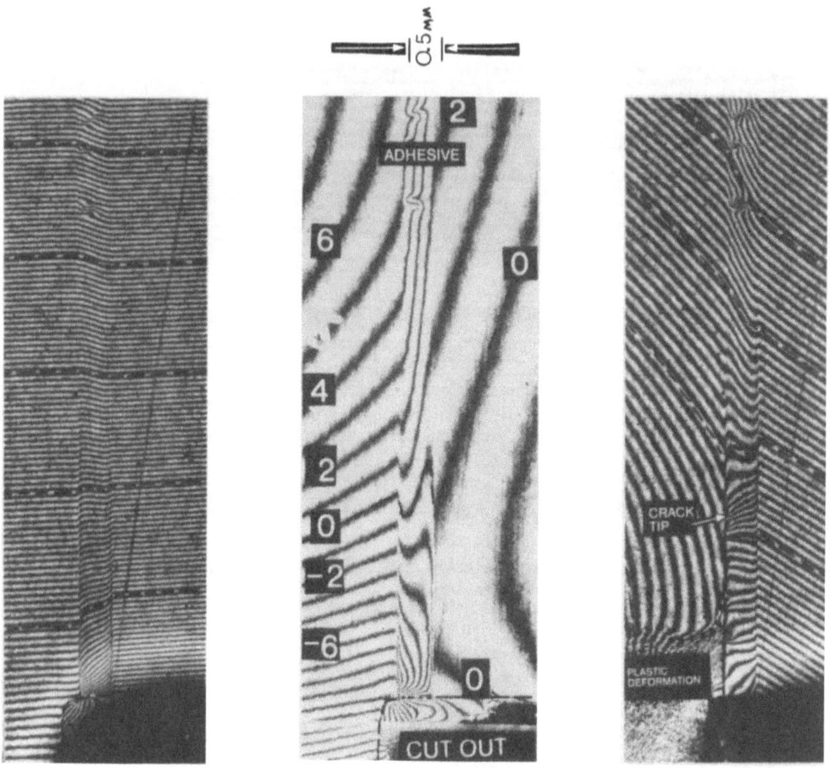

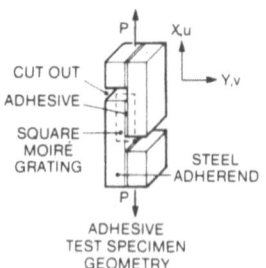

a) 3000N   b) 3750N   c) 4750N

Figure 7   v-moiré displacement fringes close to the cut out of a

TAST joint as a function of load.

In the present case, at the centre of the lap along CC' (see Figure 8) the quantities $\frac{\partial v}{\partial x}$, $\frac{\partial u}{\partial x}$ are small compared with $\frac{\partial u}{\partial y}$. Also ε_x and ε_y are small compared with $\frac{\partial u}{\partial y}$ and with unity. Consequently

$$\gamma_{xy} \simeq \Delta u / \Delta y = \frac{1}{f} \frac{\Delta Nx}{\Delta y}$$

Also the peel strains can be approximated as

$$\varepsilon_y = \frac{\Delta v}{\Delta y} = \frac{1}{f} \frac{\Delta Ny}{\Delta y}.$$

where ΔN is the fringe count across the adhesive of thickness Δy.

Figure 9 shows the number of moiré fringes, corrected for zero load across the centre of the bond, against the average shear stress on the bond. This is linear up to 4000 N with a slope of 1.02 GPa which is equivalent to the apparent shear modulus of the adhesive as would be measured using a zero gauge length extensometer positioned at the centre of the bond. The shear strains at the centre are lower than the average value along the bond length and taking this into account the modulus of the adhesive measured using this technique is 925 MPa.

The shear deformations through the thickness of the adhesive at the cente of the bond C-C', and at cutout A-A' for a load of 4000 N are shown in Figure 10. These were obtained by measuring the distance between adjacent fringes. The shear strains are uniform through the thickness of the adhesive at the centre of the bond and they are also approximately uniform at the cutout. The shear strains reach a maximum at the cutouts.

The peel strains are compressive at the central region of the joint and tensile at the cutouts. In Figure 7b the maximum fringe gradient $\frac{\partial u}{\partial y}$ occurs at the cutout and it is clear that there are high localised peel strains at the interfacial region close to the continuous adherend.

For loads greater than 4000 N a fringe discontinuity at the interface in both u and v displacement fields is observed and marks the presence of an interfacial crack which grows from B' towards the joint centre as the load is increased. Figure 7c illustrates this effect for the v fringes at 4750 N. A similar crack from A growing towards C is observed at the other

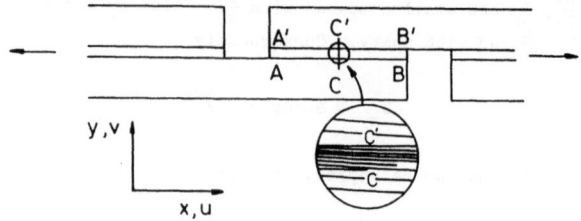

Figure 8   TAST reference joint.

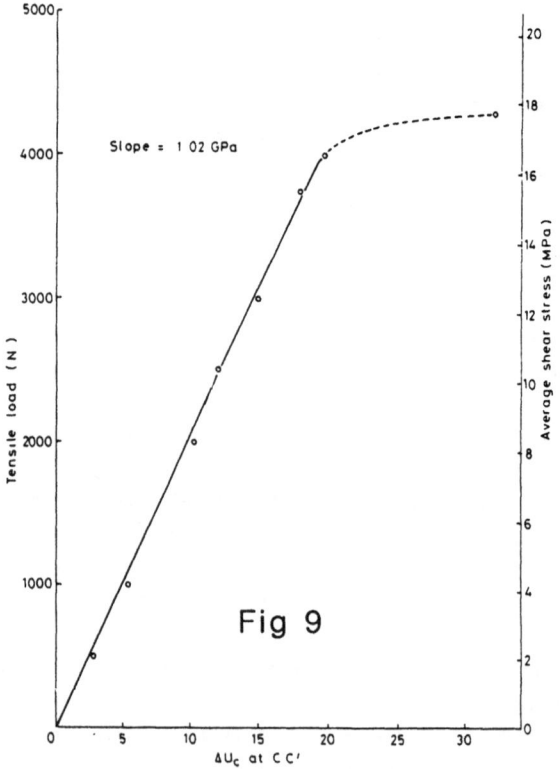

Slope = 1 02 GPa

Fig 9

ΔU$_c$ AT THE CENTRE OF THE BOND ACROSS CC'
AGAINST AVERAGE SHEAR STRESS

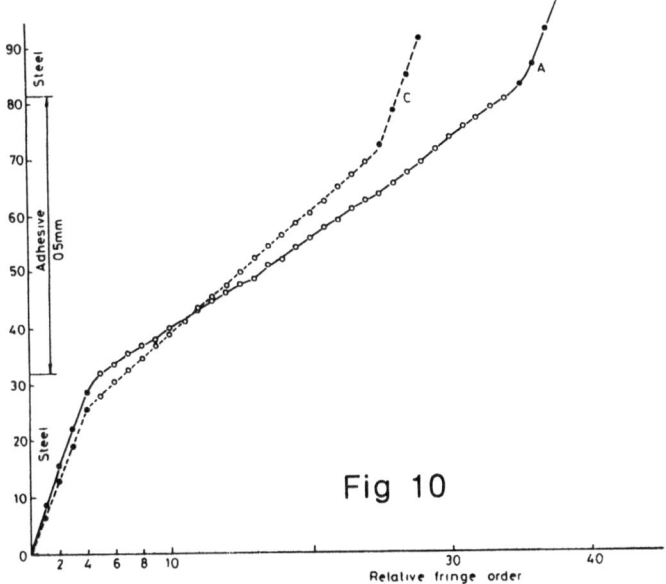

ΔU AGAINST Y IN THE ADHESIVE ALONG LINES AA′ & C′C FOR P = 4000N

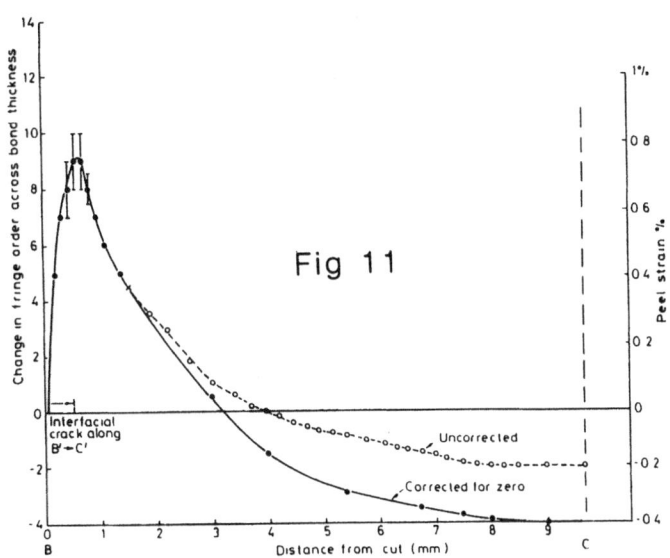

V - PEEL DEFORMATIONS ACROSS THE BOND THICKNESS AS A
FUNCTION OF DISTANCE FROM THE CUT FOR P = 4300N

cutout. The initiation of these cracks correspond to the start of the non-linearity observed in the shear strain results in Figure 9. The effective bond length of the joint carrying the load is reduced and for a given load the shear strains carried in the uncracked central region of the adhesive are increased accordingly. The maximum peel strain averaged across the bond is observed at the crack tip. The magnitude of this strain remains effectively constant at ∿ 0.8%. The results taken from a specimen loaded to 4300 N are given in Figure 11. The curve labelled uncorrected is obtained directly from fringe information taken from the loaded specimen. The corrected curve is obtained by taking into account the v fringes observed in the sample when it was unloaded from 4300 N. Plastic deformation of the steel adherends at the cutouts is also observed and this is clearly seen in Figure 7c in the steel adjacent to the crack.

Quantitative data taken from the specimen held at 3750 N and before interfacial cracking occurred is shown in Figures 12 and 13 for the average peel strain and shear strain distributions along the bond length. These data indicate shear strains of 2% and peel strains of 0.75% at the ends of the joint. Shear strains and peel strains are a minimum at the joint centre at 1.42% and −0.25% respectively. The results predicted by the BISEPS-LOCO Harwell computer codes are also included. The shear strain distribution along the adhesive length measured using the moiré techniques fits closely with that predicted using non-linear continuum mechanics models developed at Harwell. The average peel strain distribution along the bond length also fits the predicted distribution well. However it is clear from the moiré results that the peel strain distribution through the adhesive thickness is non-uniform particularly near the cutouts. These high localised peel strains are of major importance in controlling the initiation and growth of the cracks in the adhesive bond and its ultimate failure. By observing the initiation and growth of the cracking in such adhesively bonded samples the failure processes are being more fully understood. This in turn will enable the more accurate modelling and prediction of adhesive bond strengths and give a greater confidence in structural adhesive joint design procedures.

CONCLUSIONS

- Laser moiré interferometry has been shown to be a unique and invaluable technique in the study of full field deformations in adhesive joints.

- The technique has been successfully applied to directly measure the stiffness and strength characteristics of a structural epoxide adhesive.

- The shear strains are uniform through the thickness of the bond.

- The shear strains are non-uniform along the length of a shear bond being a minimum at the centre of the bond and reaching a maximum at the joint ends. There is good agreement with the predicted strain distributions measured using non-linear continuum mechanics analysis programs under development at Harwell.

- The equivalent peel strains are compressive at the centre of the joint and tensile at the ends. Here also the theoretical predictions fit the experimental data well when the avarage through thickness strains are used. However, moiré interferometry indicates that the peel strain is non-uniform through the thickness of the joint and this is especially marked close to the joint ends.

- It is the highly localised peel strains close to the interface near the joint ends which initiate initial failure of the adhesive. However, the failure is by a stable crack growth with incresing load. The maximum peel strain at the crack tip is approximately constant at 0.8%. The corresponding shear strains are supported by the uncracked adhesive.

- Laser moiré interferometry is fulfilling its potential and is and will continue to contribute superb and exciting advances beyond our current insight.

236

ACKNOWLEDGEMENTS

The author wishes to acknowledge the most generous help of Dan Post of VPI
in supplying master gratings and for his advice in setting up the laser
moiré technique. Thanks are also due to Graham Barnes for the construction
of the loading frame used in this work. The development of laser moiré
techniques was undertaken as part of the Underlying Research Programme of
the UKAEA.

REFERENCES

1. R.B. Krieger, paper Proc. 22nd Nat. SAMPE Symposium, San Diego CA, USA,
   April 1977.

2. W. Althof, Progress in Structural Bonding through Effective Bond Line
   Stress Analysis, SAMPE Vol. 22, p. 784-95, 1977.

3. D. Post, Moiré Interferometry at VPI and SU Experimental Mechanics,
   Vol. 23, No. 2, June 1983, p. 203-210.

4. P.S. Theocaris, Moiré Fringes in Strain Analysis, Pergamon Press,
   New York, 1969.

5. E.M. Weissman and D. Post, Moiré interferometry near the theoretical
   limit, Applied Optics 21(9), pp. 1621-1623, 1982.

6. D. Post, Moiré Interferometry, Chap. 7, Handbook of Experimental
   Mechanics, A.S. Kobayashi, Editor, Prentice-Hall, Englewood Cliffs NJ
   (1986).

15

THE SELECTION OF ADHESIVES FOR AIRCRAFT REPAIRS BASED ON FUNDAMENTAL MECHANICAL PROPERTIES OF THE RESINS THEMSELVES (PHASE II)

By K.B.ARMSTRONG
Senior Development Engineer Airframe

British Airways plc
P.O.Box 10
Heathrow Airport (London)
Hounslow TW6 2JA

INTRODUCTION

The first part of this project (1) concentrated on the measurement of mechanical properties of a number of adhesives. Tensile strength, tensile modulus and elongation to failure were obtained and related to the lap shear strength of joints using Aluminium Alloy test pieces.

Diffusion and Solubility Coefficients were also obtained using distilled water at room temperature.

This second part of the work was directed primarily at the selection of adhesives for composite repairs. Initially, six materials were included but as the programme progressed a number of other adhesives and some composite matrix resins were added. With these latter, the tests were of more limited range and directed towards elucidating particular questions. It is interesting that work on these lines was recommended as long ago as 1976 by Patrick (11).

It was also decided to carry out compression testing but without an extensometer in addition to the tests which had been used earlier. Materials tested included many of those used in the first phase as well as those used for the first time.

At a later stage wedge tests were carried out to try to relate fracture energy to lap shear strength.

237

A significant amount of manufacturer's data for tensile strength and a small amount for compression strength were obtained. This data revealed that British Airways tensile and compression tests, without an extensometer, gave very low results for tensile modulus but that compression modulus values were much nearer the published figures.

It was found possible, by plotting British Airways tensile modulus values against Manufacturers' ASTM-D638 values, to obtain a curve that allowed estimation of the tensile modulus of other adhesives corresponding to the standard method. Fig.1. It was then clear that cold-setting resins for composite repairs have moduli (at room temperature) quite close to those of hot-curing resins normally used for composite manufacture.

Unfortunately, at the time of writing, there was not enough compression modulus data available from Manufacturers to plot a similar curve against British Airways compression modulus data.

Fig:1 allowed a number of values, from Reference 1, to be re-plotted against true modulus values and enabled a direct comparison with the work of Palmer (2) on composite matrix resins. It seems that cold-setting resins may be just as acceptable as hot-setting resins for composite manufacture or repair, except where high temperature performance is required.

MATERIALS, TESTING PROGRAMME AND TECHNIQUES

The adhesive materials used initially were:-

Redux 408	Two part, Epoxy adhesive
Redux 410NA	Two part, Epoxy adhesive
Redux 501	Two part, Epoxy matrix resin
Hysol EA9309.3NA	Two part, Epoxy adhesive
3M-EC9323	Two part, Epoxy adhesive
Bostik 5435/TM2	Two part, Acrylic adhesive (pre-mix type)

Redux 408 and 501 were offered by Ciba-Geigy as their materials which came closest to meeting the requirements of a British Airways Specification EE-R76-1(A) for Cold-Setting Repair Adhesives.

It was decided to use the terms "adhesive" and "matrix resin" because the mechanical and physical properties indicated a clear distinction between the two in most cases and because the work of Palmer (2) showed that the properties of a composite fall off rapidly when the modulus of the matrix falls below 450,000 psi.

A "matrix resin" for the purposes of this paper is defined as "A resin fluid enough to wet fabrics easily and having a tensile modulus in excess of 450,000 psi, giving a rigid laminate but having a relatively low shear strength in the lap shear test".

An "adhesive" is defined as "A paste of moderate viscosity, having a tensile modulus of 450,000 psi or less and giving a high shear strength in the lap shear test."

The differences between matrix resins and adhesives will be discussed in more detail later. Inevitably some materials are borderline between the two and Redux 308A, in particular, can fulfil either role.

Tensile test pieces were moulded from each of the adhesives and resins listed above. Only one mould was made and therefore each specimen was an individual resin mix. A few brittle ones were lost on removal from the mould. The brittleness of Permabond E37, Redux 408 and Redux 501 forced the use of Molykote 33, a silicone grease, as a mould release agent. Aerosol spray, non-silicone types were not good enough. Manufacture was spread over about 9 months. Materials were weighed out on electronic scales to 0.1 gramme and great care was taken to work to that accuracy because for some materials even an error of 0.1 gramme was a significant percentage error. The usual quantity mixed was about 25-30 grammes.

It is worth noting that resin systems of moderate, non-splashing, viscosity are inherently safer to work with than those where the base resin is very stiff and the curing agent water thin. These require care and patience until they begin to mix and there are two problems with accuracy of mixing. Firstly, the viscous resin tends to stick to the container and secondly, there is the danger of losing the liquid curing agent by splashing. Wherever possible it is desirable to formulate a two-part mix where both parts have a moderate and roughly equal non-splash viscosity. Hysol EA9330 and Redux 410NA typify desirable examples.

A further addition to the previous work was that more specimens were made for each material than in the previous programme and tested after post cure at 50°C for 3 hours and at 80°C for 1 hour to study the effect of post-curing on mechanical properties. 3 specimens were made for each cure temperature, except for Bostik 5435/TM2, which was only cured at room temperature.

For all the above, except Bostik 5435/TM2, 3 more specimens were made and cured at room temperature. These were immersed in distilled water. The Redux 410NA, Hysol EA9309.3NA and 3M-EC9323 specimens were tested after approximately 2, 4 and 6 months immersion. Because of their brittleness all three Redux 408 and 501 specimens were tested after 6 months immersion because it was felt that the scatter in the results due to brittleness might mask the effect of water immersion.

As the programme went along some more resins were added. The reason for this was that the work of Palmer (2) showed that many adhesives had a modulus too low for them to be used effectively as matrix resins and it was the intention of this second phase of the work to look at resins for composite repairs. In fact, a programme of testing of flexural and compression specimens had been planned to elucidate the properties required of matrix resins. This would have run in parallel with the programme to elucidate the properties required of an adhesive to make a strong lap joint. Fortunately, the discovery of the paper by Palmer (2) made the composite testing unnecessary.

These additional resins were all "matrix resins" except Permabond E38 which is classified as an "adhesive". They were tested as follows:-

Permabond E37   RT cured samples only

Permabond E38   60°C cured samples only

Shell Epikote 815 + Epikure RTU   90°C post cure tested dry & wet
                                       120°C post cure tested dry
                                       A large number of RT cured samples
                                       made and tested at different
                                       strain rates.

Shell Epikote 828 + RTU   100°C post cure for comparison
                                       with Manufacturer's data
                                       60°C post cure
                                       RT cure tested dry
                                       RT cure tested wet

Fiber Resin Corp FR7020   RT cure tested dry
                                       RT cure tested wet
                                       55°C post cure
                                       80°C post cure

Diffusion and Solubility Coefficents were obtained for all the systems used in this work. See Table 2A.

Glass transition temperature Tg was obtained, both dry and wet, for each tensile test material and cure temperature by Dr.W.W.Wright, of R.A.E. Farnborough. See Table 1A.

Testing of all the specimens showed that in almost every case failure occurred through bubbles. Where bubbles could be seen by holding the specimens up to a bright light the positions were marked, holes drilled and filled with the next mix of adhesive. However, this technique, although generally successful, only served to reduce the size of bubble through which failure occurred. Small bubbles could not be seen and so could not be repaired.

Tensile testing proved difficult in two ways:-

1. It was found that the cast adhesive materials were not adequately gripped by the serrated wedge grips fitted to the RDP/Howden No EU500BS testing machine. 320 grade Aluminium Oxide paper folded over the ends of the specimens was found sufficient to provide an adequate grip.

2. Having loaded the tensile specimen into the lower grip and taken care to achieve the best possible alignment, it was found that on lowering the upper grip the specimen buckled. This occurred because the tapered wedges came down a little further after some load was applied as the abrasive paper gripped the specimen.

The buckling that resulted, fractured one of the brittle specimens and caused the design of "sideplates". The following technique proved successful. Sideplates were made from pieces of Aluminium Alloy, approximately 50mm wide, 3mm thick and shorter than the specimens by the length of one of the ends required to be held in the grips. Abrasive paper was bonded to one face of each sideplate with Redux 410NA. The specimen was fitted in the lower grip and carefully aligned. Next the upper grip was locked open and lowered to a position found by experiment to be suitable. The sideplates were then positioned carefully to stand on the lower grip and inside the open upper grip. The upper grip was then released to clamp the sideplates and through them to grip the specimen. This technique was completely successful but the presence of the sideplates prevented observation of the specimen while under load. This did not matter in the case of brittle specimens because they failed suddenly and without warning. Tougher specimens were tested as follows without sideplates. Each specimen was located in the bottom grip, complete with abrasive paper as before. The upper grip was then released but held open by hand and allowed to progressively clamp onto the upper end of the specimen with the abrasive paper in place. As this caused progressive buckling of the specimen the grip was kept open by hand to prevent the buckle becoming excessive and the machine started. As the slack was taken up by the machine the grip was allowed to fully close on the specimen. This always occurred with some buckling remaining and the specimen finally came under pure tensile loading as it straightened out.

The advantage of this technique was that some toughened adhesives, in particular Hysol EA9309.3NA and 3M-EC9323, could be seen to craze in several places other than the point of failure. In some cases bubbles could be seen to elongate before failure.

The toughest adhesive of all those tested, Bostik 5435/TM2 acrylic, was also sensitive to bubble size. The bubbles were so big that they could easily be measured with a small magnifier. A plot of failing load v bubble size showed a clear relationship and a calculation of stress x bubble diameter showed that even such a tough material followed Griffiths' equation. No wonder the brittle materials were sensitive to bubbles and flaws too small to see.

A number of materials were tested in tension after various periods of water immersion. The only difference in technique was that these specimens were removed from storage in distilled water, wiped dry and weighed immediately. They were then taken to the testing machine and tested within 30 minutes of removal from the water.

Because of the extreme brittleness and notch sensitivity of Redux 408 and 501 at room temperature, a little blending seemed worth trying in order to improve these properties. It was decided to mix some resin and hardener of Redux 408 (and similarly of Redux 501) and then into each mixture to add some mixed Redux 410NA, a much tougher resin. The procedure was to prepare 100 parts of mixed Redux 408 (or 501) and add 20 parts of mixed Redux 410NA and similarly for 40 parts of mixed Redux 410NA. This gave four different compositions in each of which the mixed Redux 410NA was found compatible with mixed 408 and 501. Each sample was allowed to stand for 10 minutes before pouring. Each of the blends was poured into the mould and allowed to outgas without a covering release sheet.

Later in the programme it was decided to carry out wedge tests to try to relate fracture energy to lap shear strength. Wedge specimens were all made using chromic acid anodised 2024-T3 Aluminium Alloy and using zero bond pressure. Paste adhesive glue lines were controlled (approximately) using .010" shims and film adhesives were laid down, the two halves pressed together by hand and then cured in an oven. Further blends of Redux 410NA with Redux 408 and Redux 501 were made for these tests and fracture energy was plotted logarithmically in Fig:2A and on a linear scale in Fig:2B. Fracture Energy v Blending Ratio is shown in Fig:26. Wedge tests were also carried out on EC2216, AF163-2K, Hysol EA9330, Hysol EA9309.3NA, Hysol EA9321, Redux 308A, Epikote 815 + Epikure RTU, Epikote 828 + Epikure RTU and FR7020. Wedges were driven in at 10mm/minute using the RDP/Howden machine and taking advice given by (9) Joneja et al. This was successful and critical to producing the fairly low scatter in results that was achieved. The specimens were stabilised by hand until the wedges were sufficiently far in to be self-supporting. The force required to drive the wedge, although variable, was related to the fracture energy of the adhesives tested.

At a late stage in the water uptake tests carried out to obtain Diffusion and Solubility Coefficients, it was decided to add a strip of 7075-T6 unclad Aluminium Alloy to each jar and to note if any corrosion occurred. This was done because one of the adhesives in the first programme had caused corrosion in service on some repairs. pH and conductivity of the water in each jar were measured periodically.

A very recent paper (12) tends to confirm the possibility of an effect of pH or resin chemistry when it says, "It was shown that the ability of surface treatment to provide joint durability is dependent on the adhesive selected, supporting the belief that joint durability is determined by the interface comprising the aluminium surface and the adhesive."

Blocks of resin were also cast for each of 19 resin systems and 3 specimens of each material 10mm square by 25mm were machined and tested to ASTM-D695. Compression testing was generally very successful and almost all specimens reached 10% plastic strain without failure. The curves varied considerably. In some cases a clear ultimate value was achieved, others showed a constant plateau and the softer ones flattened and the load continued to rise. The results are shown in Fig:5.

In the case of Epikote 828 + RTU a high ultimate value was reached but the elastic limit was rather lower than might have been expected. When selecting resins for compression application it could be important to consider that the order of merit may change depending on whether ultimate or elastic limit values are used.

Compression modulus values were generally lower than the Manufacturers. Insufficient data was available to allow British Airways' tests to be compared with the Manufacturers.

Some difficulty with the compression tests was found when attempting to make specimens of Redux 308A and AF163 film adhesives. Blocks were made by pressing together about 50 layers of film. Curing in an oven, without pressure, produced a dense foam bearing no relation to the normal cured film. A further attempt was made using vacuum pressure. A more dense foam was produced but the results are marked "suspect" because the values seem too low compared with the two-part pastes.

The foaming suggests that film adhesives contain more volatiles than is generally admitted.

The two-part pastes also contained a few bubbles but test results were generally very consistent. Results would be expected to be more consistent for compression tests than for tensile tests because:

a)   The three samples in each case were made from one block and therefore only one mix was involved.
b)   Compression properties are less sensitive to flaws and bubbles.

RESULTS AND DISCUSSION

Tensile Tests

Tensile tests were carried out on all the resins used in this programme and in each case values for tensile strength, tensile modulus, elongation at failure and strain energy at failure were obtained. In many cases these tests were also carried out after the complete test piece had been immersed in distilled water for periods from 2 to 6 months.

In general it was found that tensile modulus was little affected by post-curing. Some resins actually reached their highest modulus from a RT cure and values after post-cure were slightly lower. Redux 408 showed a slight increase in modulus after a 50°C post-cure. A 90°C post-cure was found to be beneficial to the tensile strength of Epikote 815 + Epikure RTU but a 120°C post cure reduced it almost to the RT cure level. For Epikote 828 + RTU both 60°C and 100°C post-cures improved the tensile strength. At a constant modulus improved tensile strength goes hand-in-hand with improved elongation. The tensile elongation of Redux 501 is shown in Fig:21. These results are also reflected in Strain Energy at failure for Redux 501 in Fig:24.

Electron microscope studies of the effect of heat treatment on microstructure could be helpful to assist in developing optimum cure cycles. In doing this it must also be remembered that 80°C in the sun or other temperatures for engine cowlings and during supersonic flight may be met in service and some unintentional heat treatment or cure cycle can occur.

The results from the Phase II Programme (room temperature cure data only) were added to figures from (1) and in the case of tensile modulus they were corrected using Fig:1 (Tensile Modulus ASTM D-638 v Tensile Modulus in B.A. Test without an Extensometer). The figures from (1) were re-drawn in the light of additional data and a further attempt was made to relate lap shear strength to the various mechanical properties of the resin systems tested.

A study of Figs:3, 4, 6, 7 and 9 shows that Epikote 815 + RTU and Epikote 828 + RTU do not show up as well as one might expect. Fig:3 shows some correlation of resin tensile strength with lap shear strength but beyond about 4,000 psi this drops off for some resins although the hot-cured AF163 and Redux 308A do better. Fig:4 shows a correlation with tensile modulus but this falls off beyond 350,000 psi. Fig:6 shows

quite a good correlation of elongation to failure with lap shear
strength but it is not good for Redux 501 blended with Redux 410NA or
for Epikote 815 + RTU, or Epikote 828 + RTU. A function, (tensile
strength $\div \sqrt{G}$) was used to represent a function of the Volkersen Stress
Concentration Factor. In Fig:7 this function was plotted against lap
shear strength in an attempt to get an order of merit of joint strength
related to resin strength and this Stress Concentration Factor. Again,
it can be seen that composite matrix resins (usually brittle compared
with adhesives) fall out of the pattern for adhesives used for making
joints. Finally, in Fig:9 it can be clearly seen that again, most of
the composite matrix resins have a high strain energy at failure but a
poor lap shear strength. Consideration of all these figures suggested
that although, up to a point, a correlation can be made between the
fundamental properties of tensile strength, tensile modulus and
elongation to failure and lap shear strength there must be another
factor or other factors not yet taken into account.

Effects of Blending Redux 410NA With Redux 408 And Redux 501

Diagrams were plotted for the various mechanical properties from 100%
Redux 408 or Redux 501 to 100% Redux 410NA to enable more optimum blends
to be tried if this experiment proved successful. See Figures:18-26.
The results from very limited data indicate that "Rule of Mixture"
behaviour applies quite well for compression properties. The tensile
strength of Redux 501 was significantly improved but the blending made a
smaller difference to 408. Another interesting finding with Redux 501
was that a post cure at 80°C gave a similar tensile strength improvement
to toughening with 20% of Redux 410NA. The amount of Redux 410NA
required to improve tensile strength did not reduce modulus values very
much.

A study of Figs:18, 21 and 24 revealed that tensile strength,
elongation to failure and strain energy at failure all go together as
would be expected. However, Fig.25 showed very little improvement in
lap shear strength by blending in small amounts of Redux 410NA with
Redux 408 or Redux 501. This was particularly interesting because
Epikote 815 + RTU and Epikote 828 + RTU also have quite good tensile
strengths, elongation to failure and strain energy at failure but in
spite of these desirable characteristics they develop poor lap shear
strength.

It was therefore considered that the missing factor, required to
define the properties of a good adhesive, might well be fracture energy
and to check this possibility a considerable amount of wedge testing was
undertaken.

Wedge Tests

Wedge testing was carried out on most of the adhesives used in this
programme and a few used in the previous programme. Also included were
a number of additional blends of Redux 410NA with Redux 408 and 501 to
provide graphs of blending ratio v wedge test fracture energy extending
from 100% 410NA to 100% 408 or 501. See Fig:26.

Low fracture energy, low elongation to failure materials fit the straight line parts of Fig:2A and Fig:2B. It would seem that only those materials with a high fracture energy and a good elongation to failure can reach the higher parts of the curves. Materials with a low fracture energy but high tensile strength and elongation to failure do fit the curves (e.g. Epikote 815 + RTU). In Figures 2A and 2B it is two of the high fracture energy materials, EC2216 and EA9330 that do not fit the curve. EC2216 has a very high elongation and low tensile strength and modulus. EA9330 has a lower modulus than Redux 410NA.

When fracture energy is low the joint is likely to fail when the tensile strength of the resin is first exceeded at the ends of the joint and the local stress cannot be relieved by plastic strain, crazing or cracking.

Experiment has shown that in some cases, when testing tough adhesives a "ping" is heard as the "spew" fails at the ends of a joint but this is usually only just before failure. The load does not increase much above this point before failure occurs. Lap joints made with brittle adhesives fail with one bang and not in two or more stages. It would seem that brittle adhesives, with a high modulus, fail because the high stress at the ends of the joint cannot be relieved by plastic strain or crack growth because they have no plastic strain and the critical crack length is too short. On the other hand, tough adhesives are of lower modulus, exhibit some plastic strain and can tolerate a longer crack because of their higher fracture energy. Adams & Wake (13) have shown that failure of brittle adhesives can be predicted by a maximum principal stress criterion whereas a maximum principal strain criterion was found more suitable for toughened adhesives. The crazing seen in EC9323 and EA9309.3NA before failure would suggest that this is the case. Low strength, low modulus adhesives probably fail due to stress rather than strain.

Figs:2A and 2B indicate that some optimum combination of tensile modulus, tensile strength, elongation and fracture energy is required to give a high lap shear strength. Observation of lap joints under test showed an increasing rotation of the joint with increasing load indicating that some peel strength or fracture energy must be necessary.

The test pieces that failed at the highest loads suffered permanent deformation indicating that the yield point of 2024-T3 Aluminium Alloy had been exceeded. Adhesively bonded joints tend to fail when the yield strength of the adherend has been exceeded because of the additional strain produced in the adhesive layer. Some of the adhesives that caused permanent deformation in the lap shear test also caused permanent deformation in the wedge test. The fracture energy values obtained are probably too high because of this. Crack growth in water was probably too low for the same reason.

Fig:2C shows a considerable drop in fracture energy after 10 days immersion in filtered water at room temperature. The value of fracture energy at which the curves level out would seem to be more important than the original dry values when comparing adhesives.

The wedge test really brought out the difference between "adhesives" and "matrix resins". The former were always at the higher end of the

fracture energy scale and the latter always at the lower. In terms of
fracture energy after a period of time dry, and especially after a
period of time immersed in water, the difference between "flexibilised"
resins and "toughened" resins could also be seen. Crack growth for
"flexibilised" resins was much greater. When splitting specimens to
check the mode of failure it was found that a sharp hammer blow could
change even EC2216 from a tough to a brittle failure mode. All of the
adhesives tested behaved in this way except the toughest, Redux 410NA
and 3M-AF163-2K. This confirmed that driving the wedge in slowly in a
controlled manner was the right approach. Adhesives for use in
situations of impact loading need to be tested at appropriate loading
rates to match the end use conditions.

It was also observed that although most of the failures were
cohesive even with brittle materials, Redux 501 gave totally apparent
adhesive failure. A blend of three parts Redux 501 and one part Redux
410NA gave apparent adhesive failure for two thin glue lines and totally
cohesive failure for one thick glueline.

As the proportion of tough Redux 410NA increased the proportion of
cohesive failure increased although all specimens were made from the
same sheet of Aluminium Alloy, anodised and bonded on the same day. It
would seem that adhesive properties as well as good surface preparation
can affect the achievement of cohesive failures. It was also noted
that, within the limits recorded, a thicker glue line gave a higher
fracture energy. The results of wedge tests are shown in Figs:2A, 2B
and 2C and also Fig:26. The work of Mall & Johnson is of interest as
they have studied the effect of fracture energy in great detail (15),
(16).

Lap Shear Tests

A number of lap shear tests were also carried out to ASTM D-1002 on
specimens made from the same variety of blends of Redux 410NA with Redux
408 and 501 used for the wedge tests described above. These tests were
made to enable lap shear strength v fracture energy to be plotted.

Water Uptake Tests

Water uptake tests using 40mm x 30mm specimens cut from the ends of
tensile test pieces as in (1) were run for all the adhesives tested.
This work was carried out as in (1) for the following reasons:

1. Because water uptake was considered likely to relate to joint
   durability.
2. Because Tg and mechanical properties were expected to fall with
   increasing water content.
3. To obtain wet Tg values.
4. To obtain Diffusion Coefficients.
5. To obtain Solubility Coefficients.

Corrosion tests have shown that corrosion is most likely to be
caused by high pH due to the specific ions present and can occur at very
low water uptakes. However, high water uptake is of interest because it
is likely to lead to debonding whether or not corrosion is involved.

Diffusion and Solubility Coefficients obtained are listed in Table 2A. This Table shows that quite significant differences exist among the materials tested in both Diffusion and Solubility Coefficients and the data obtained is helpful to the selection of adhesives. The Diffusion Coefficient for Redux 775 cured at 100 psi is similar to the epoxies but the Diffusion Coefficient when cured at zero psi is higher by an order of magnitude. The Solubility Coefficients are very high compared with epoxies and make the durable performance of joints, made with Redux 775, difficult to explain. It is hoped that the analysis of the water in each jar will help to clarify the situation.

Corrosion Tests

These tests were carried out on all the samples previously subjected to tensile testing. The results obtained after 130 days immersion were very interesting. It was found that no corrosion occurred until the pH of the water rose above 7. Conductivity and pH both rose with time and in most cases passed a peak after corrosion occurred and then began to fall. Peak values of pH are shown in Fig:8. Redux 408 and 410NA both contain an inhibitor and caused no corrosion even at a pH of just above 7. Redux 308A produced a pH above 7 but no corrosion in 130 days. Bostik 5435/TM2 gave no corrosion and after 130 days the pH was well below 7. FR7020 was odd in that samples cured at RT and 50°C did not cause corrosion in 130 days but the sample cured at 80°C did cause corrosion.

All other samples caused corrosion of varying severity. This work will be fully reported when analyses of the water in each jar have been completed by R.A.E. Farnborough.

At the present time, there is a move to take Chromate inhibitors out of adhesives on health grounds. Unless the risk is significant it would seem to be more important to ensure that they are included in all resins used for Aluminium Alloy bonding. Fortunately research is already underway to develop safer corrosion inhibitors and one such programme is reported in (14) by Matienzo, Shaffer, Moshier and Davis.

In view of the rapid corrosion of untreated Aluminium Alloy the need for either Chromic or Phosphoric Acid anodising can be seen to ensure good durability as well as bond strength. Where a surface preparation of lower corrosion resistance than anodising is used it is even more important to use corrosion inhibited, low water uptake adhesives. Two other factors should be reported in this context:

1. Control samples of 7075-T6 Unclad, in distilled water only, corroded even faster than those immersed with an adhesive. The rise in pH was very rapid but the conductivity rose very little. Once severe corrosion had occurred the pH fell equally rapidly. After 44 days the pH was back to 7.2. The initial pH of the water at the time of adding the Aluminium Alloy was only 5.1.

2. It was suggested by Dr.Lees that the cheap soda lime glass bottles used for these tests might cause some rise in pH themselves. Two jars were filled with distilled water only and the pH monitored. It was confirmed that some alkali is leached

from the glass alone as the pH did rise with time. It is therefore recommended that in any future work on these lines either "Pyrex" glass jars should be used or other containers which will not themselves affect the results.

The above mentioned testing was carried out because the operating temperature of composites is limited by:

(a) The glass transition temperature of the resin.
(b) The water uptake of the resin in service.

Tg is affected by the choice of curing system and usually, but not always, by the temperature of cure.

Water uptake is dependent on the choice of curing agent and the degree of cure.

The in-service water uptake of a composite matrix resin eventually reaches about 50% of the value obtained at saturation after immersion in water at room temperature. Tg falls by about 20°C for every 1% of water absorbed [Ref.8, Wright]. Table 2A suggests that the reduction in Tg for cold-setting resins is much less than this.

The importance of resin Tg is much greater for repairs to areas of composite skin than it is for metal lap joints or joints in composite parts. For joints it can be seen from Fig:4 (Lap Shear Strength v Resin Modulus) that a considerable loss of modulus makes little difference to the joint strength and therefore short times under load, where creep is not a problem, are likely to be acceptable. However, for composites the compression strength of a skin has a direct relation to modulus below about 450,000 psi. It is, therefore, important to avoid loading composites at temperatures where the resin modulus falls below this figure.

The problem of selecting appropriate cold-setting resins for composite repairs is that the original parts are cured, most commonly at 180°C and often at 120°C in order to retain adequate hot/wet performance at 80°C.

The hottest air temperature in the world seldom exceeds 50°C and therefore even at the extreme an aircraft surface would be likely to have cooled to 50°C by the end of the take-off run. Problems arise in two particular circumstances:

(1) Aerodynamic heating of the surfaces of supersonic aircraft when the structure is under load.
(2) High surface temperatures (80°C) on upper surfaces when parked in the sun.

These conditions have to take into account the level of moisture uptake likely to have occurred in the resin matrix and the effect this has on Tg and the composite mechanical properties.

A resin for use as a composite matrix therefore requires the properties detailed by Palmer in (2) plus a Tg appropriate to the maximum in-service temperature in spite of it's moisture content. A low water uptake will, therefore, be very helpful to achieving this.

For non-structural composite parts, lightly loaded parts and parts that are not loaded until an aircraft is airborne, a Tg of 50°C at 50% of the resin saturation water content would seem to be reasonable and Tg values of 40°C or lower would probably be acceptable. Fortunately, the vast majority of the flight time of modern airliners is spent at high altitude where the air temperature can be as low as -55°C and on rare occasions -76°C. As a result of a small amount of kinetic heating skin temperatures are often around -25°C. One resin used for bonded metal repairs very successfully, except for long term corrosion, has a dry Tg of only 11°C. It has also been used successfully for a very lightly loaded composite repair. This item is only loaded in flight.

For structural parts of supersonic aircraft, a Tg at 50% water content appropriate to the aircraft service temperature must be achieved. For lightly loaded parts the mechanical properties of the resin at the cruise temperature are probably more important than Tg.

Composite structural parts of subsonic aircraft that carry significant load, while heated by direct radiation from the sun, need to be made with resins having a Tg around 80°C at 50% of their saturation water content. These resins, whether cold-setting or not will be post-cured at 80°C in service. Repairs can be made with resins of lower performance than the original provided that extra layers of tape or fabric are used to restore the original compression strength, (1) Armstrong and Fig:17.

It will be seen from Figures 1 - 7 and 9 and 11 that metal bonding adhesives, in addition to composite matrix resins, can be more readily compared on the basis of fundamental resin mechanical properties than by comparison of lap shear tests alone. However, it must be said that lap shear testing should continue because lap joints are the most common joints. It would seem that the lap shear test does measure fracture toughness, though not quantitatively, because only those adhesives with a good fracture energy achieve the highest lap shear performance.

From the data presented in (1) and the updated and additional tables and figures of this paper it should now be possible for Engineers to tell their Chemist colleagues, fairly accurately, the mechanical and physical properties required of adhesives and matrix resins for specific end uses. It is not suggested that these properties will be easy to achieve.

CONCLUSIONS

This programme of work has proved to be particularly interesting and its results are considered to suggest guidelines for the development of improved adhesives and composite matrix resins. It has shown that the performance of adhesives in lap joints can be related to their fundamental mechanical properties in the same way as Palmer (2) has shown that the performance of the matrix resin governs the performance of a composite using a particular fibre. The value of toughened matrix resins to improve damage tolerance of composites is detailed in (10). The work also shows that a better comparison of adhesives can be made on the basis of fundamental mechanical properties than by a comparison of strength in a lap joint. Specific conclusions are listed below:-

1. Moisture uptake affects the modulus and Tg of a resin more than other properties. For most materials tested it had little effect on tensile strength. FR7020 and Epikote 815 + RTU showed a significant loss of tensile strength and tensile elongation.

2. Post-curing is most likely to affect tensile strength. It may improve Tg and reduce water uptake slightly.

3. Blending a tougher resin with a brittle one can considerably improve tensile strength. The affect on moduli is likely to be in line with the "Rule of Mixtures". However, the amount of "tough" resin required to improve tensile strength is fairly small and the optimum amount only reduces modulus values slightly.

4. Tg values need to be supplied by manufacturers. Ideally, they should be obtained using the Torsion Pendulum Test to ASTM-D-2236 rather than Differential Scanning Calorimetry (DSC). DSC requires expert interpretation and standard runs usually go a long way above the actual Tg. This means that any re-run on the same sample is measuring a new material which has been:

    (a) Dried to a lower water content.
    (b) Cured to a higher degree than the sample from which it was cut.

    However, DSC is quicker, cheaper and can produce results from smaller samples of material.

5. The three major problems with cold-setting resins are:

    (a) Low initial Tg because of choice of chemistry and low temperature of cure.
    (b) Loss of Tg due to water uptake.
    (c) The tendency of many of them to cause corrosion unless inhibitors are incorporated.

6. Further work needs to be done to find methods of raising the Tg of cold-setting resins, either from a room temperature cure or by warm post-curing.

7. Development of cold-setting resins could usefully concentrate on achieving a low water uptake without loss of the required mechanical properties.

8. The preliminary work mentioned in this report suggests that the durability of adhesive bonds to Aluminium Alloy and probably other metals could be significantly dependent on the pH of the resin and/or any extracts from the resin by leaching with water or other fluids. The effect of pH on the bonding of composites and composite joint durability could be a useful study. All adhesives likely to cause corrosion should contain suitable inhibitors. These may not be the same for all metals.

Where composites are used to repair metal parts the effects of water uptake and pH are likely to be more severe because water can penetrate the entire composite surface. The pH needs to be suitable for both the metal and the composite in this case.

The need for this was confirmed when some composite repairs to Aluminium Alloy components using a particular resin did suffer corrosion. This adhesive also produced corrosion in the tests mentioned in this paper.

9. The optimum adhesive modulus for a bonded joint, Fig:4, is a little lower than the optimum modulus for making a composite. This suggests that if composite parts are assembled by co-curing, they should use a layer of toughened film adhesive to make the joints and not the same resin as that from which the composite itself is made. This could be particularly useful when composites are made using high modulus brittle resins.

10. Brittle resins, of high modulus, tend to have a low tensile strength because of a low fracture energy and high sensitivity to flaws and bubbles.

11. Compression strengths seem to vary from about the same as tensile strength to about 1.7 x tensile strength.

12. Data for adhesives and composite matrix resins should, ideally, be provided in graphical form across the entire temperature range as typified by Fig:27. All temperature dependent properties could usefully be presented in this way. This data indicates that the fall in properties with temperature is seldom sudden and that some useful load can still be carried for short periods at temperatures a little above Tg. A repeat of these curves using material at 50% of its saturation water uptake would help to indicate the performance of composites after a long period of outdoor service.

13. Good lap shear strength seems to require an optimum combination of resin tensile strength, modulus, elongation and in particular fracture energy. Further study in this area should be helpful. From the limited data available "T" peel strength would appear to require a similar combination.

14. When repairing a metal part with a composite patch it is desirable to bond the first layer of fabric or tape to the metal with a corrosion resistant film or paste adhesive. This serves two purposes:-

    a) The patch is bonded with a tough, joint making adhesive.
    b) A corrosion inhibited adhesive will reduce the probability of metal corrosion. The remaining layers of composite should be bonded with a high modulus matrix resin to give a rigid laminate.

15. The properties of adhesives for good lap joints on Aluminium Alloy and matrix resins for composite materials are listed below. Because the moduli of different metals and fibre reinforcements vary considerably, the values for adhesives may need modification for metals other than Aluminium Alloys and Composites other than Carbon-Fibre.

Lap Joint Adhesive Properties

A good lap joint resin for Aluminium Alloy should have the following properties. Optimum values may be different for other metals or composite materials.

Tensile Strength: 4,000 to 7,000 psi

Tensile Modulus: 350,000 - 500,000 psi

Tensile Elongation: Between 3 and 7%

Water Uptake: 4%

Tg sufficient to give 1000 psi lap shear strength at maximum service temperature.

pH of water soluble extract between 6.0 and 7

Fracture Energy in Wedge Test: 1 KJ/m to ASTM D-3762.

If greater shock or impact resistance is required then lower values of Tensile Strength and Modulus and higher values of elongation and fracture energy are likely to be beneficial

More attention needs to be paid to pH to give greater durability

Corrosion inhibitors should be included whenever it is necessary to do so.

The chemistry of the resin may need to be chosen to suit the metal being bonded.

Composite Resin Properties from Ref.2 Palmer

A good composite resin for carbon-fibre composites should have the following properties to give the best combination of impact and mechanical properties.

Tensile Strength: Between 8,000 and 10,000 psi

Tensile Modulus: Above 450,000 psi

Tensile Elongation: Between 5 and 6%

Water Uptake: 4%

Tg 40°C above maximum service temperature when dry to allow for reduction due to water uptake. For cold-setting resins 20°C above maximum service temperature may be sufficient.

Low viscosity to ensure good penetration and wetting

Fracture Energy in Wedge Test: 0.5 KJ/m to ASTM D-3762.

Increasing elongation to failure reduces impact damage area but lowers mechanical properties. Higher fracture energies are desirable where impact resistance is required

Optimum values may be different for other fibre systems

REFERENCES

1. Armstrong, K.B., Properties of Adhesives for Composite and Bonded Metal Repairs. Adhesion 11, Elsevier-Applied Science 1986, pp118-174.

2. Palmer, R.J., Investigation of the Effect of Resin Material on Impact Damage to Graphite/Epoxy Composites. McDonnell Douglas Corporation, Douglas Aircraft Company, Long Beach, California 90846, NASA Contractor Report 165677.

3. Berry, D.B.S., Buck, B.I., Cornwell, A. and Phillips, L.N., Handbook of Resin Properties, Part A - Cast Resins. Prepared and published by Yarsley Testing Laboratories, The Street, Ashtead, Surrey, Ministry of Defence Contract No. K/LT32B/1932 October 1975.

4. Swartz, Charles A., A Novel, Damage Tolerant, Toughened Epoxy Resin, 31st International S.A.M.P.E. Symposium, April 7-10 1986, pp 163-176.

5. Chaudhari, M., A New High Performance Epoxy Resin for Advanced Composites, 31st International S.A.M.P.E. Symposium, April 7-10 1986, pp 563-570.

6. White, W.D., Resin-Hardener Systems For Resin Transfer Moulding, 31st Internation S.A.M.P.E. Symposium, April 7-10 1986, pp 622-634.

7. Hewitt, Ralph W., Schlaudt, Laurie M., and Bonner David C., Use of New Epoxy Resin Systems For Wet Filament Wound, High Performance Structures, 31st International S.A.M.P.E. Symposium, April 7-10 1986, pp141-152.

8. Wright, W.W., A Review of the Influence of Absorbed Moisture on the Properties of Composite Materials Based on Epoxy Resins, Royal Aircraft Establishment Technical Memo MAT 324.

9. Joneja, Surendra J. and Newaz, Golam M., Evaluating SMC Bonds Using a Wedge Test, published Adhesives Age October 1985 pp 18-22.

10. Damage Tolerant Composites, Aerospace Engineering, December 1987, pp 8-11.

11. Structural Adhesives With Emphasis On Aerospace Applications. Treatise on Adhesion & Adhesives Vol 4, Edited R.L.Patrick. Published Marcel Dekker 1976.

12. A Comparative Study of Aluminium Joint Durability With Varying Surface Treatments And Adhesives, by Hamill J.L, Furlong S.L and Emptage M.R. Published Society for the Advancement of Material & Process Engineering (S.A.M.P.E), 19th Technical Conference, October 13-15 1987, Crystal City, Virginia, USA.

13. Structural Adhesive Joints In Engineering, Adams R.D and Wake W.C. Elsevier Applied Science Publishers 1984.

14. Environmental And Adhesive Durability of Aluminium-Polymer Systems
     Protected With Organic Corrosion Inhibitors, Matienzo L.J, Shaffer
     D.K., Moshier W.C., Davis G.D. Journal of Materials Science 21
     (1986), pp1601-1608.

15. Characterisation of Mode 1 And Mixed-Mode Failure of Adhesive Bonds
     Between Composite Adherends, Mall S., and Johnson W.S, S.A.M.P.E
     Journal May/June 1988, pp322-334.

16. A Fracture Mechanics Approach For Designing Adhesively Bonded
     Joints, Johnson W.S. and Mall S., NASA TM 85694, National
     Aeronautics and Space Administration, September 1983.

Acknowledgements

The encouragement of Mr.K.W.Allen, of The City University and the
invaluable help of Dr.W.W.Wright of the Royal Aircraft Establishment,
Farnborough, is gratefully acknowledged.

TABLE 1
Tg Dry & Wet After Various Cure Temperatures

ADHESIVE	Tg DRY & WET AFTER VARIOUS CURE TEMPERATURES					
	After RT Cure Dry	After RT Cure Wet	After 50°C Dry	After 50°C Wet	After 80°C Dry	After 80°C Wet
EA 9330	11°C	NR	NR	NR	27°C	NR
EPIKOTE 815+RTU	48°C	NR	61°C	52°C	69°C	47°C
EPIKOTE 828 + VERSAMID 125	47°C	NR	51°C	37°C	47°C	NR
EC 2216	NR	NR	--	NR	NR	NR
EC 3524	NR	NR	--	64°C	NR	NR
EA 9321	--	53°C	--	47°C	NR	NR
AF 163 *	--	--	--	--	108°C	NR
EC 3559	--	50°C	--	NR	71°C	49°C
EA 9330 + 20% MICROBALLOONS	--	NR	--	NR	--	NR
EC 3568	--	NR	--	--	--	--
EC 3578	--	NR	--	NR	NR	NR
PERMABOND E34	119°C **	NR	--	NR	NR	NR

NR means tested but no satisfactory result obtained by DSC
-- means that no test carried out
* AF 163-2M film adhesive cured at 120°C for 1 hour
** figure from Permabond Adhesives
None of the wet specimens tested gave a clear cut result
The figures quoted should be viewed with caution
All data obtained on Perkin-Elmer DSC-2 at RAE Farnborough at a rate of
temperature rise of 10°C/minute. Some polymers scanned between 40°C and
280°C, others between 0°C and 280°C. Results indicate that some tests
should have been started at lower temperatures.

256

TABLE 1A
Phase II Additional Work
Tg Dry & Wet After Various Cure Temperatures

Page 1 of 2

ADHESIVE	Tg DRY & WET AFTER VARIOUS CURE TEMPERATURES					
	RT Cure Dry	RT Cure Wet	50°C Dry	50°C Wet	80°C Dry	80°C Wet
REDUX 308A *	--	--	--	--	90°C + 4	NR 4
REDUX 408	61°C	NR	56°C	51°C	NR	NR
REDUX 408 + 20% REDUX 410NA	54°C	55°C	--	--	--	--
REDUX 408 + 40% REDUX 410NA	54°C	57°C	--	--	--	--
REDUX 410NA	46°C	48°C	47°C +	52°C	49°C +	52°C
REDUX 501	60°C	50°C	65°C	58°C	NR	51°C
REDUX 501 + 20% REDUX 410NA	58°C	51°C	--	--	--	--
REDUX 501 + 40% REDUX 410NA	57°C	52°C	--	--	--	--
EC 9323	42°C	49°C	60°C	50°C	54°C	50°C
BOSTIK 5435+TM2	50°C	-1°C	--	--	--	--
EA 9309.3NA	47°C	47°C	56°C	51°C	65°C	50°C
PERMABOND E37	119°C **	61°C	--	--	--	--
PERMABOND E38	--	--	NR		--	--
EPIKOTE 815+RTU	--	--	--	--	80°C 1	NR

TABLE 1A
Phase II Additional Work
Tg Dry & Wet After Various Cure Temperatures

ADHESIVE	Tg DRY & WET AFTER VARIOUS CURE TEMPERATURES					
	RT Cure Dry	RT Cure Wet	50°C Dry	50°C Wet	80°C Dry	80°C Wet
EPIKOTE 828+RTU	93°C +	58°C	106°C + 3	NR	116°C 2	NR
FR 7020	NR	49°C	NR	46°C	NR	47°C
REDUX 775 Cured 100 psi	--	--	--	--	--	NR 4
REDUX 775 Cured Zero psi	--	--	--	--	--	NR 4

```
*     Redux 308A film adhesive cured at 170°C for 1 hour
**    Permabond result
+     Clear cut result - All other results should be treated with caution
NR    Tested but no useful result obtained
1     120°C Cure
2     100°C Cure
3     60°C  Cure
4     170°C Cure
--    Means no test carried out
```

TABLE 2
Diffusion Coefficients & Solubility Coefficients
After Various Cure Temperatures

ADHESIVE	DIFFUSION COEFFICIENT "D" m²·s SOLUBILITY COEFFICIENT "S"					
	After RT Cure "D"	After RT Cure "S"	After 50°C Cure "D"	After 50°C Cure "S"	After 80°C Cure "D"	After 80°C Cure "S"
EA 9330	1.63	13%	1.63	18% **	1.63	10.4%
EPIKOTE 815 + RTU	4.74	7.35%	3.51	5.2%	4.31	3.68%
EPIKOTE 828 + VERSAMID 125	4.43	6.25%	4.43	6.6%	4.43	7%
EC 2216	1.575	5.7%	9.07	6.0%	9.07	5.4%
EC 3524	1.194	35%	1.194	32.9%	1.194	34.4%
EA 9321	9.34	7.88%	3.46	4.89%	4.33	4.28%
AF 163 *	--	--	--	--	8.04	1.89%
EC 3559	8.81	5.15%	4.33	5.4%	6.38	4.76%
EA 9330 + 20% MICROBALLOONS	7.22	79.7%	2.83	67%	8.72	102%
EC 3568	NR	18.7%	--	--	--	--
EC 3578	NR	9.7%	NR	18.1%	NR	9.5%
PERMABOND E34	NR	1.9%	NR	1.9%	NR	2.15%

*   AF 163-2M film adhesive cured at 120°C for 1 hour
**   Suspect mix
NR   No result. These materials gave a sigmoidal uptake curve from
        which no normal diffusion coefficent could be obtained

TABLE 2A
Phase Two Additional Work

ADHESIVE OR MATRIX RESIN	DIFFUSION COEFFICIENT "D" $\frac{m^2}{s}$ SOLUBILITY COEFFICIENT "S"					
	RT Cure "D"	RT Cure "S"	50°C Post Cure "D"	50°C Post Cure "S"	80°C Post Cure "D"	80°C Post Cure "S"
REDUX 408	NR	13%	NR	12.4%	NR	7.9%
REDUX 408 +20% REDUX 410NA	3.36	11.3%	--	--	--	--
REDUX 408 +40% REDUX 410NA	4.96	8.35%	--	--	--	--
REDUX 410NA	1.38	4.8%	1.38	4.8%	1.38	4.95%
REDUX 501	6.02	14.2%	1.36	19.9%	9.92	13.4%
REDUX 501 + 20% REDUX 410NA	3.58	15.1%	--	--	--	--
REDUX 501 + 40% REDUX 410NA	3.8	15.36%	--	--	--	--
EC9323	1.67	9.1%	1.46	8.2%	1.64	6.8%
BOSTIK 5435/ TM2	NR	10.6%	--	--	--	--
REDUX 308A *	--	--	--	--	2.7	3.6%
EA9309.3NA	1.14	5%	1.14	4.6%	1.07	4.4%
PERMABOND E37	2.28	2.8%	--	--	--	--
PERMABOND E38 **	--	--	1.07	15.9%	--	--

TABLE 2A
Phase Two Additional Work

ADHESIVE OR MATRIX RESIN	DIFFUSION COEFFICIENT "D" m .s SOLUBILITY COEFFICIENT "S"					
	RT Cure "D"	RT Cure "S"	50°C Post Cure "D"	50°C Post Cure "S"	80°C Post Cure "D"	80°C Post Cure "S"
REDUX 408 #	NR	11.4%	--	--	--	--
REDUX 410 #	1.38	4.88%	--	--	--	--
REDUX 501 #	6.02	17.6%	--	--	--	--
828 + RTU	7.14	4.6%	7.14	3.6% +	6.45	2.3% ++
FR 7020	6.18	9.3%	6.18	8.8%	6.18	7.8%
REDUX 775 * 100 PSI CURE	--	--	--	--	7.36	46.4%
REDUX 775 * ZERO PSI CURE	--	--	--	--	7.59	111%
815 + RTU ***	--	--	--	--	9.26	3.4%

```
*       170°C Cure
**       60°C Cure
***     120°C Cure
+        60°C Post Cure
++      100°C Post Cure
#       Free Surface
NR      No result - A sigmoidal Curve
```

261

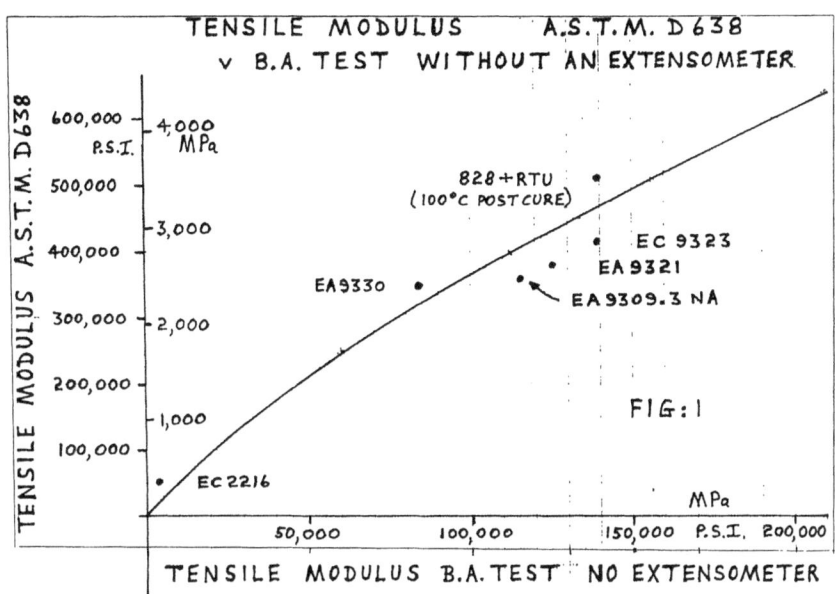

TENSILE MODULUS   A.S.T.M. D638
v B.A. TEST WITHOUT AN EXTENSOMETER

TENSILE MODULUS A.S.T.M. D638

600,000 — 4,000 MPa
P.S.I.
500,000 —
— 3,000
400,000 —

300,000 — 2,000
200,000 —
— 1,000
100,000 —

828+RTU
(100°C POST CURE)

EC 9323
EA 9321
EA 9330
EA 9309.3 NA

FIG: 1

EC 2216   MPa

50,000   100,000   150,000 P.S.I. 200,000

TENSILE MODULUS B.A. TEST NO EXTENSOMETER

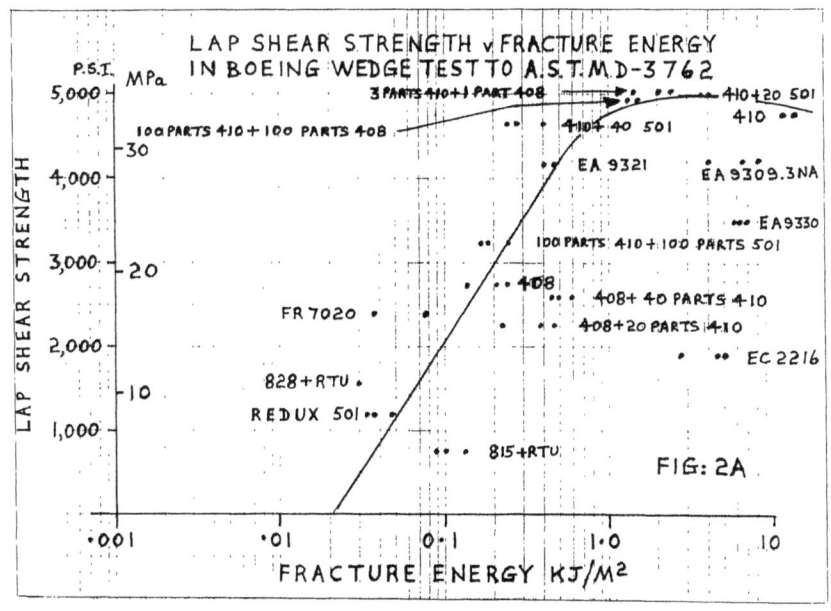

LAP SHEAR STRENGTH v FRACTURE ENERGY
IN BOEING WEDGE TEST TO A.S.T.M D-3762

P.S.I. MPa
5,000 —
— 30
4,000 —

LAP SHEAR STRENGTH

3,000 — 20

2,000 —

— 10
1,000 —

3 PARTS 410+1 PART 408   410+20 501
100 PARTS 410 + 100 PARTS 408   410
                             410+ 40 501
                             EA 9321   EA 9309.3NA
                                      EA9330
            100 PARTS 410+100 PARTS 501
      408
FR 7020   408+ 40 PARTS 410
               408+20 PARTS 410
                        EC 2216
828+RTU
REDUX 501

815+RTU

FIG: 2A

·001   ·01   0·1   1·0   10
FRACTURE ENERGY KJ/M²

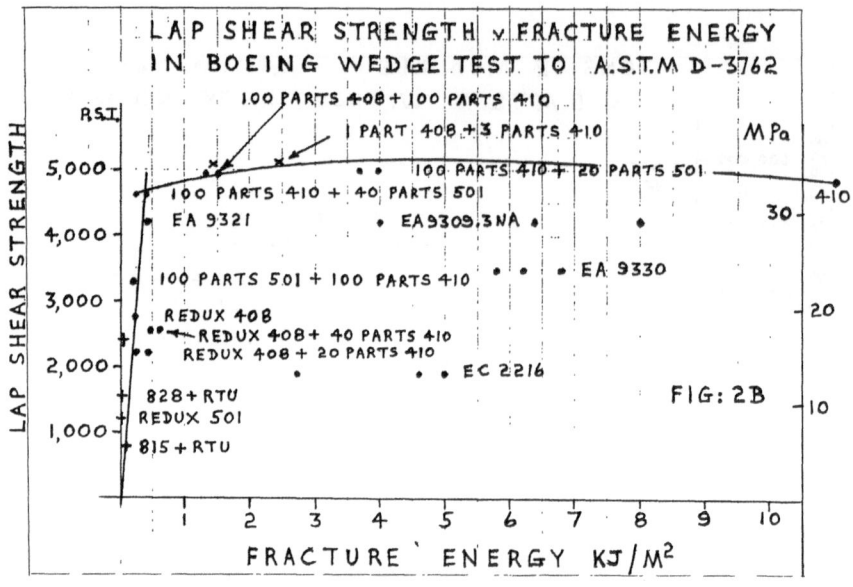

LAP SHEAR STRENGTH v FRACTURE ENERGY
IN BOEING WEDGE TEST TO A.S.T.M D-3762

FIG: 2B

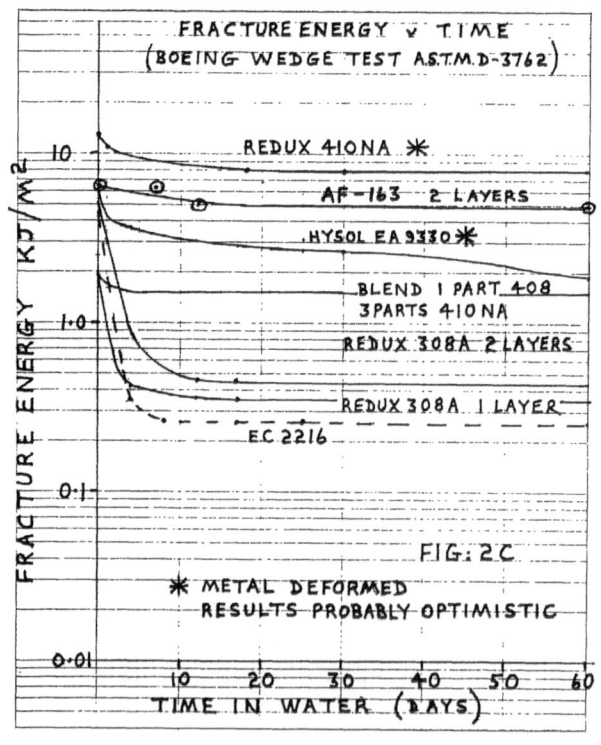

FRACTURE ENERGY v TIME
(BOEING WEDGE TEST A.S.T.M.D-3762)

FIG: 2C

* METAL DEFORMED
RESULTS PROBABLY OPTIMISTIC

263

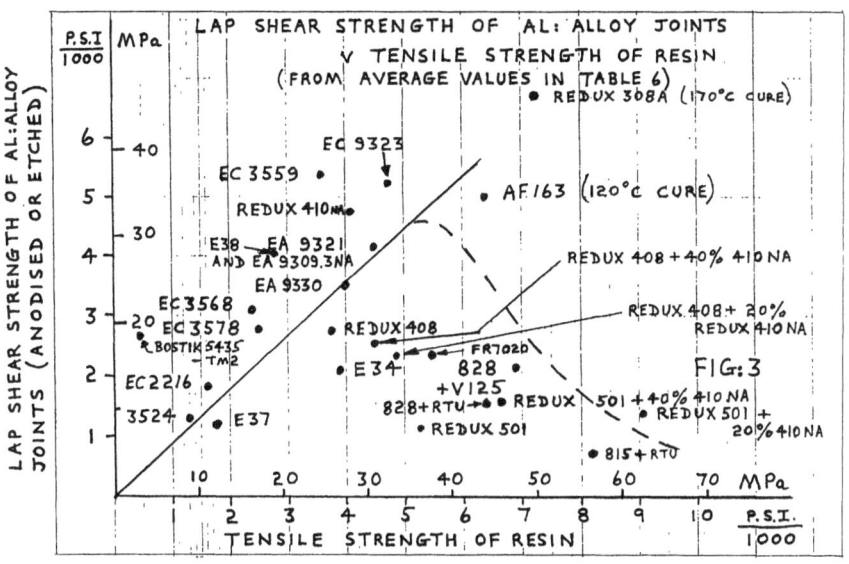

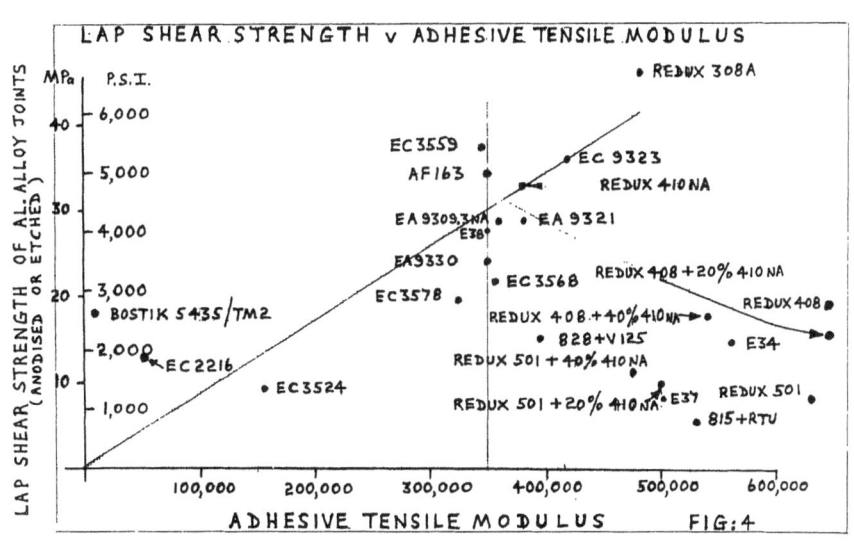

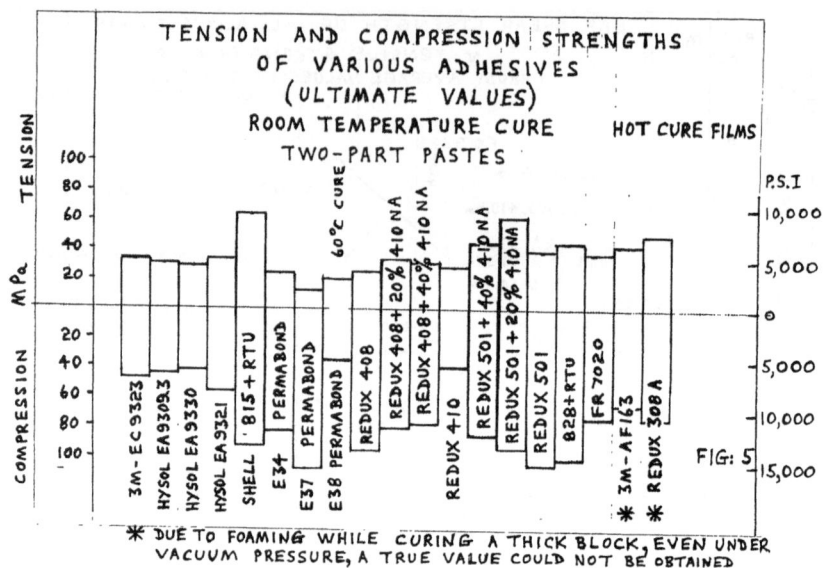

TENSION AND COMPRESSION STRENGTHS
OF VARIOUS ADHESIVES
(ULTIMATE VALUES)
ROOM TEMPERATURE CURE   HOT CURE FILMS
TWO-PART PASTES

FIG: 5

✱ DUE TO FOAMING WHILE CURING A THICK BLOCK, EVEN UNDER
VACUUM PRESSURE, A TRUE VALUE COULD NOT BE OBTAINED

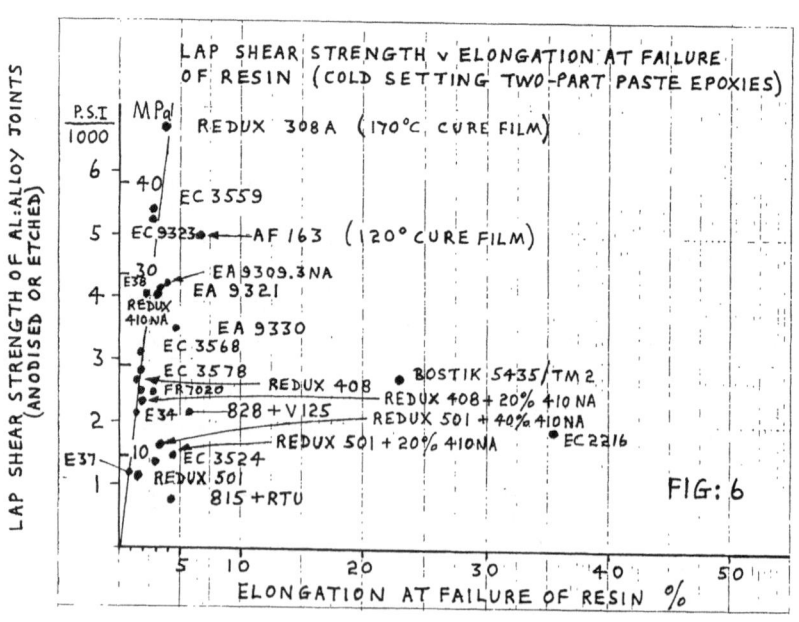

LAP SHEAR STRENGTH v ELONGATION AT FAILURE
OF RESIN (COLD SETTING TWO-PART PASTE EPOXIES)

FIG: 6

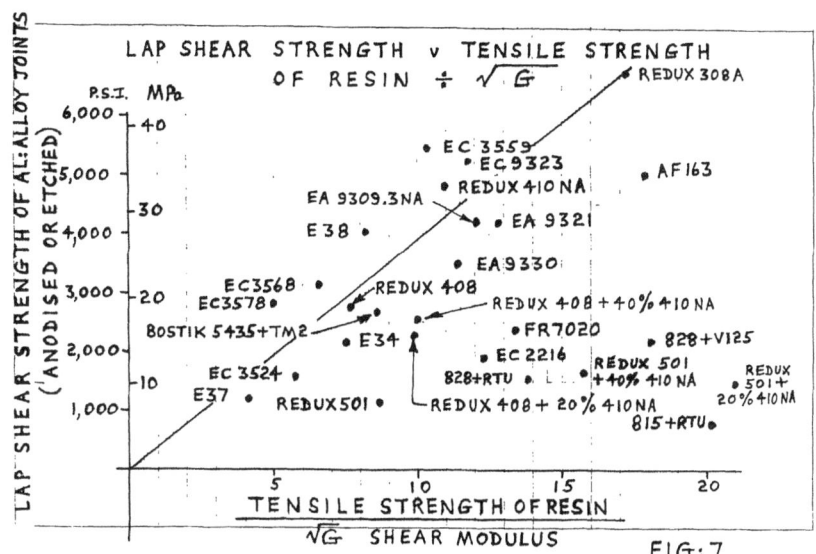

LAP SHEAR STRENGTH v TENSILE STRENGTH
OF RESIN ÷ √G

FIG:7

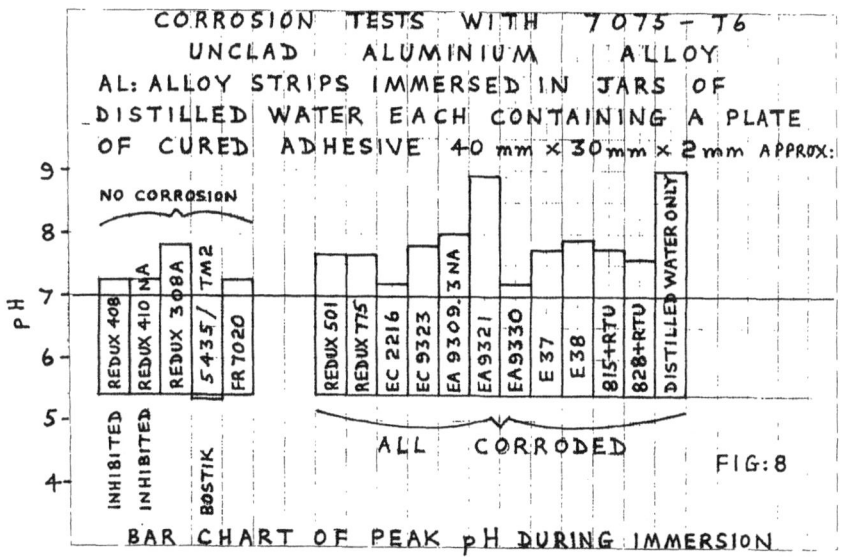

CORROSION TESTS WITH 7075-76
UNCLAD ALUMINIUM ALLOY
AL:ALLOY STRIPS IMMERSED IN JARS OF
DISTILLED WATER EACH CONTAINING A PLATE
OF CURED ADHESIVE 40mm × 30mm × 2mm APPROX:

FIG:8

BAR CHART OF PEAK pH DURING IMMERSION

266

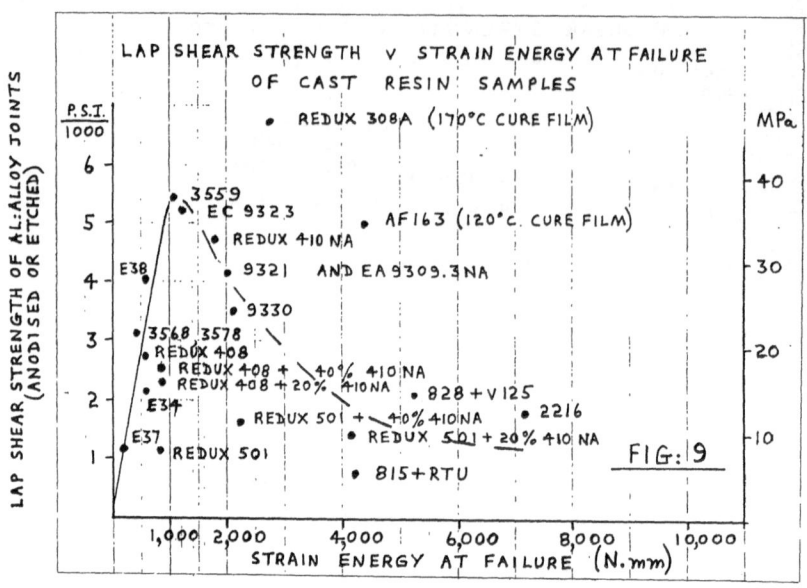

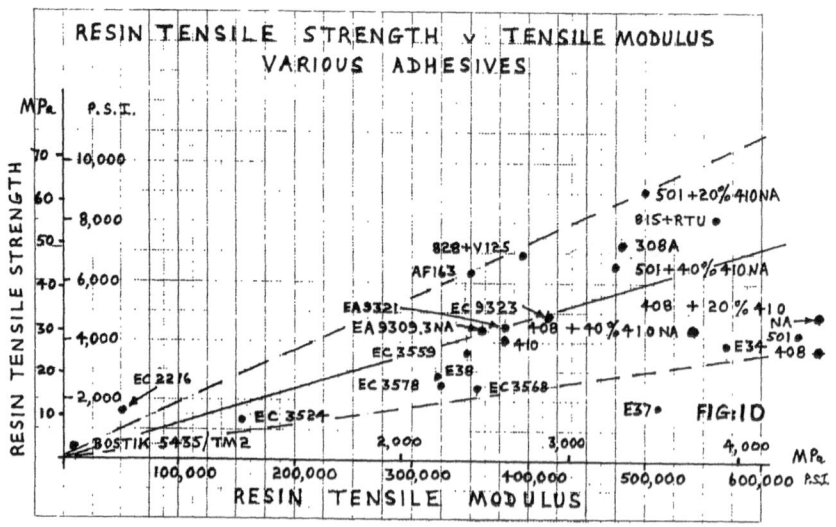

267

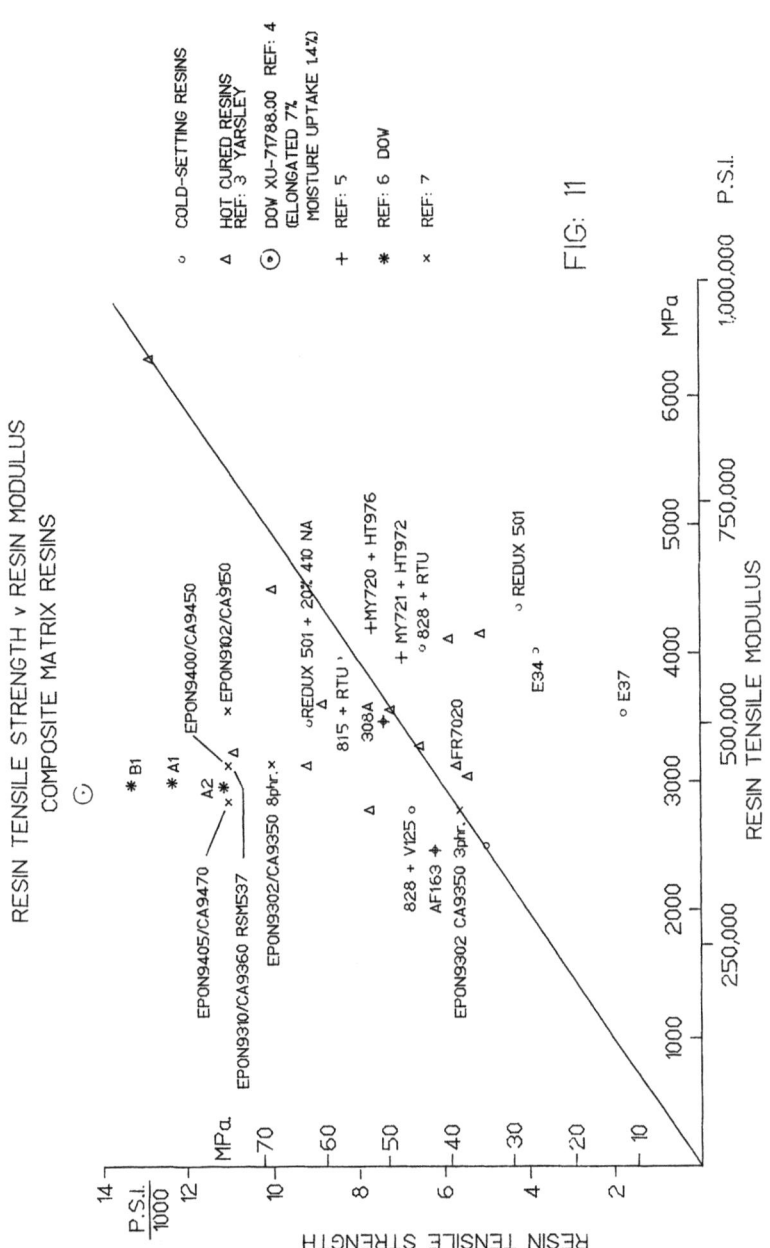

FIG: 11

268

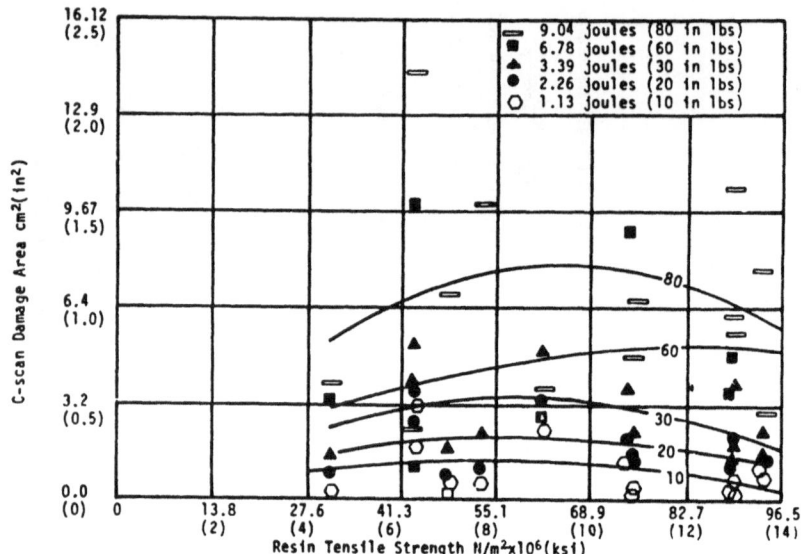

FIGURE 12. NEAT RESIN TENSILE STRENGTH VS IMPACT DAMAGE AREA
(from Ref.2)

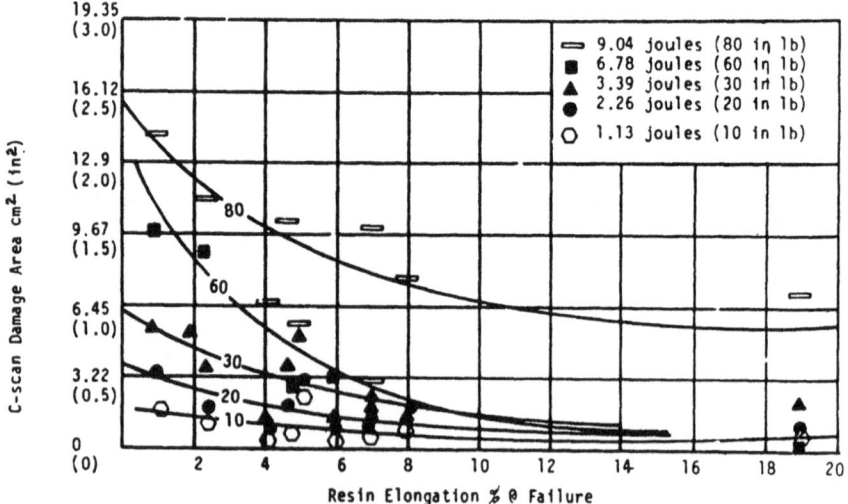

FIGURE 13. NEAT RESIN % ELONGATION AT FAILURE VS. IMPACT DAMAGE AREA
(from Ref.2)

269

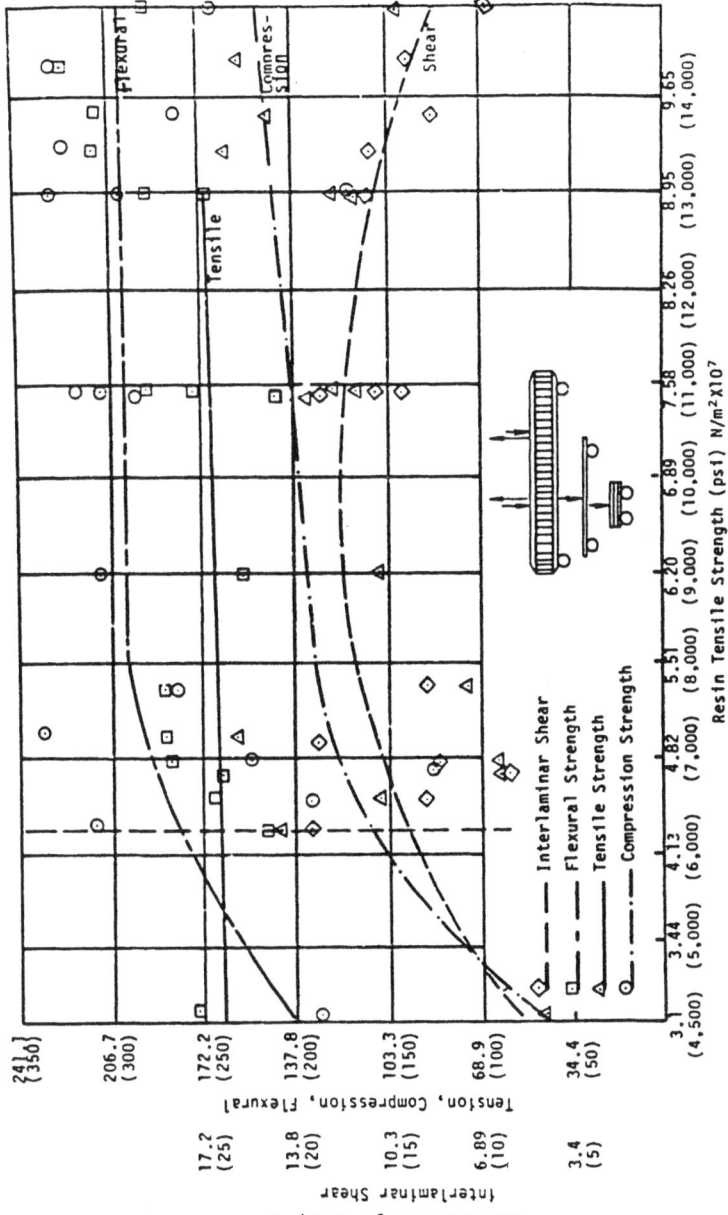

FIGURE 14. RESIN TENSILE STRENGTH VS LAMINATE MECHANICAL PROPERTIES (from Ref.2)

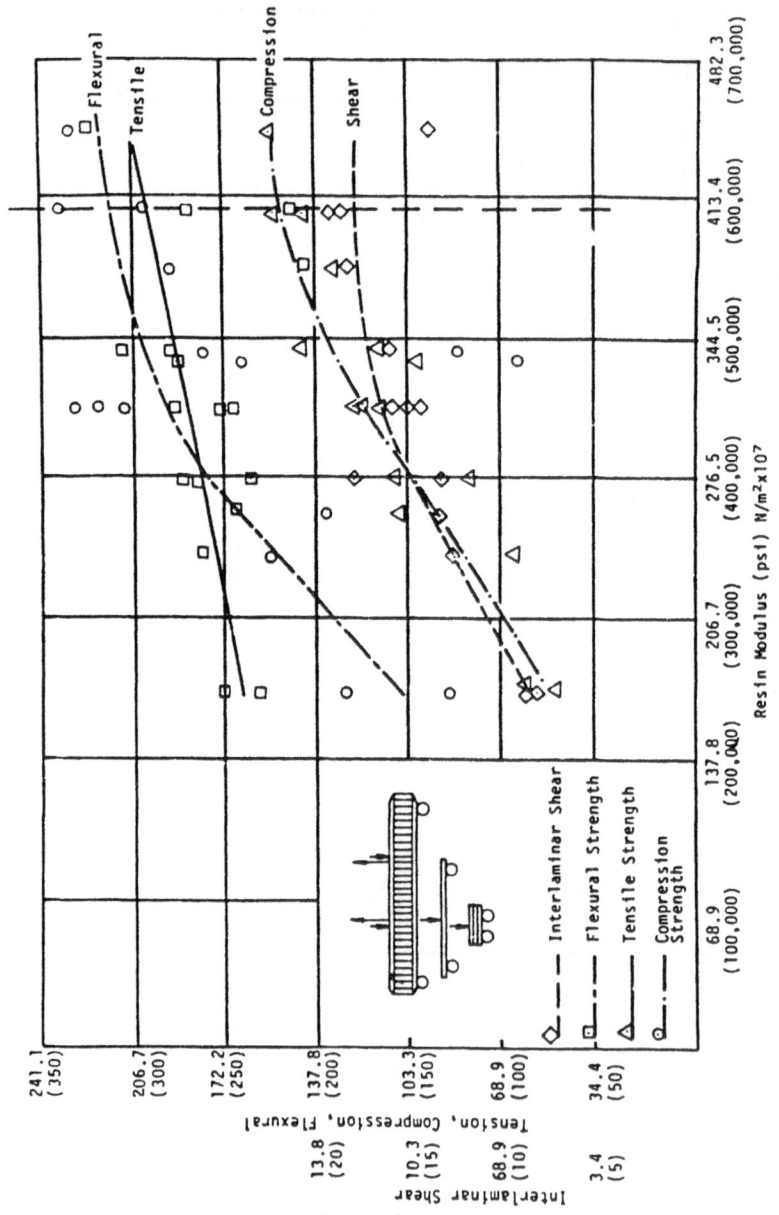

FIGURE 15. RESIN TENSILE MODULUS VS LAMINATE MECHANICAL PROPERTIES (from Ref.2)

271

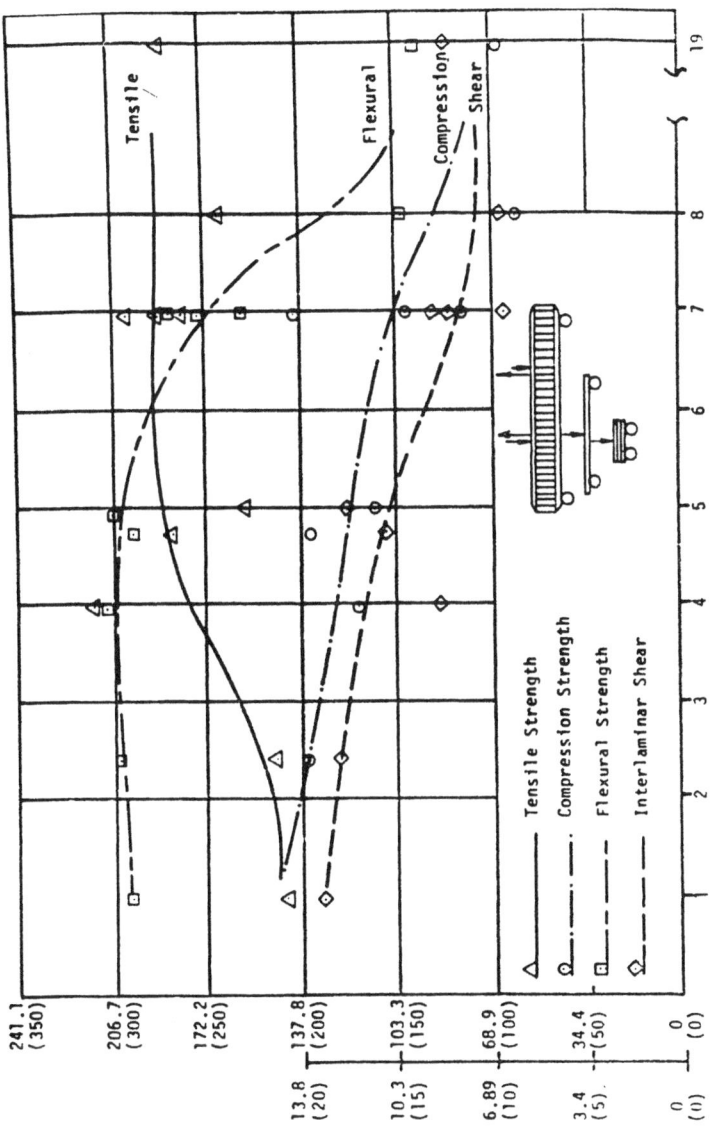

FIGURE 16. RESIN TENSILE ELONGATION VS LAMINATE MECHANICAL PROPERTIES (from Ref.2)

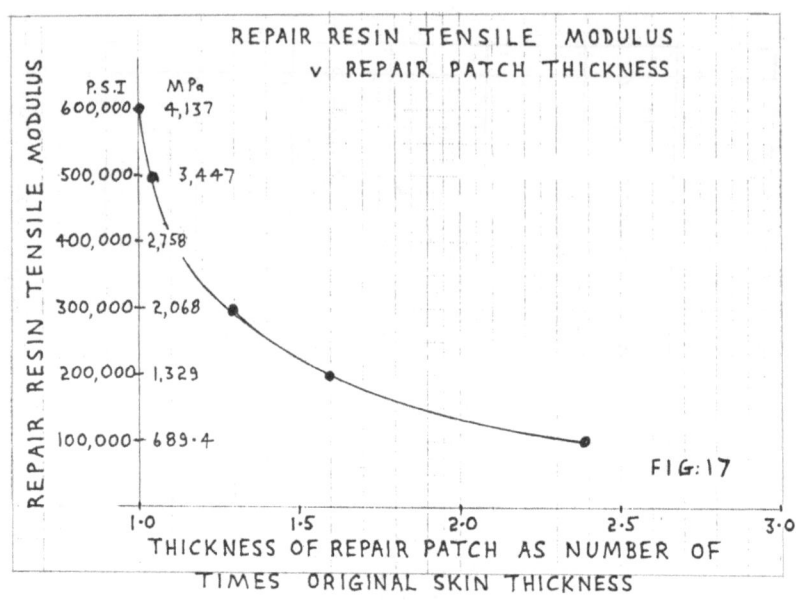

REPAIR RESIN TENSILE MODULUS
v REPAIR PATCH THICKNESS

FIG:17

THICKNESS OF REPAIR PATCH AS NUMBER OF
TIMES ORIGINAL SKIN THICKNESS

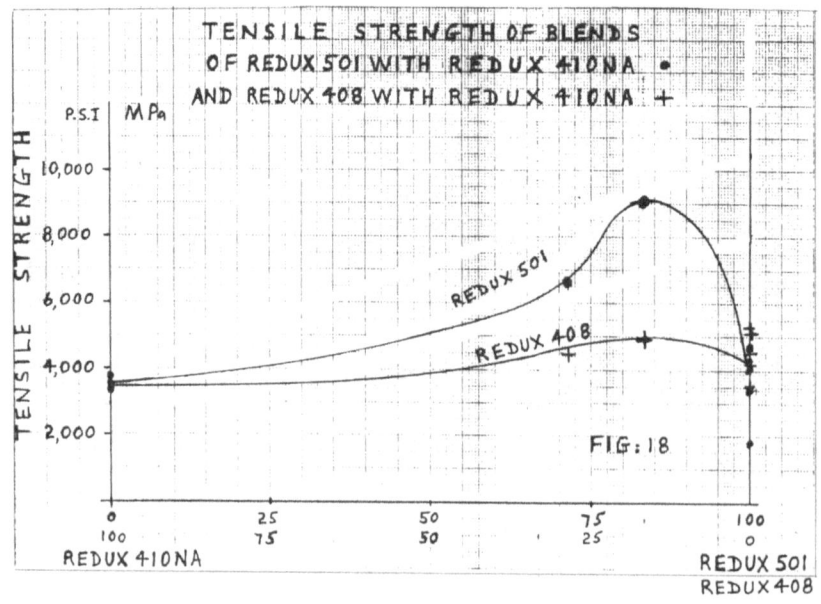

TENSILE STRENGTH OF BLENDS
OF REDUX 501 WITH REDUX 410NA •
AND REDUX 408 WITH REDUX 410NA +

FIG:18

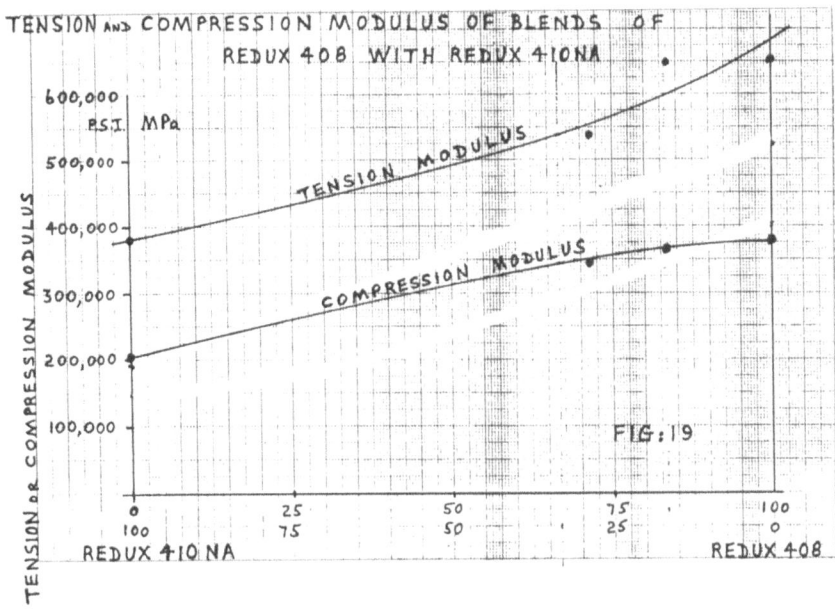

TENSION AND COMPRESSION MODULUS OF BLENDS OF
REDUX 408 WITH REDUX 410NA

FIG:19

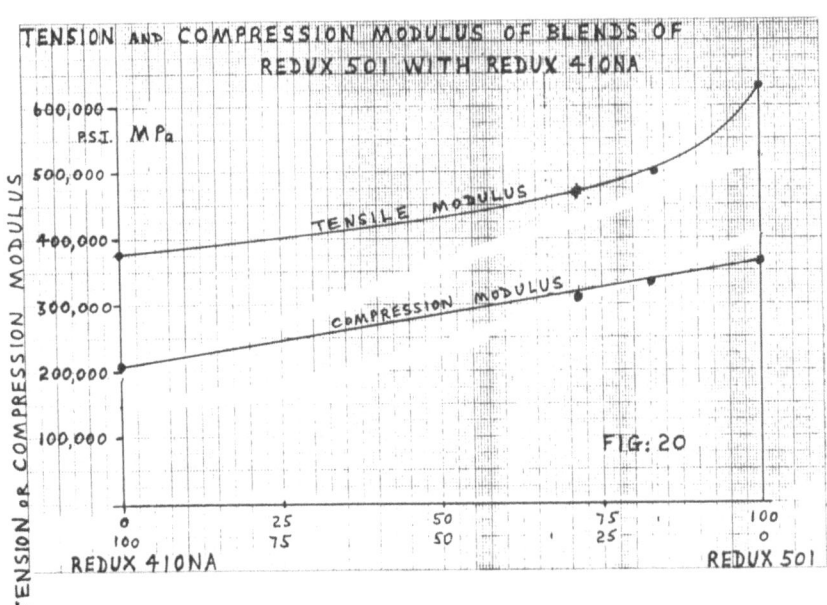

TENSION AND COMPRESSION MODULUS OF BLENDS OF
REDUX 501 WITH REDUX 410NA

FIG:20

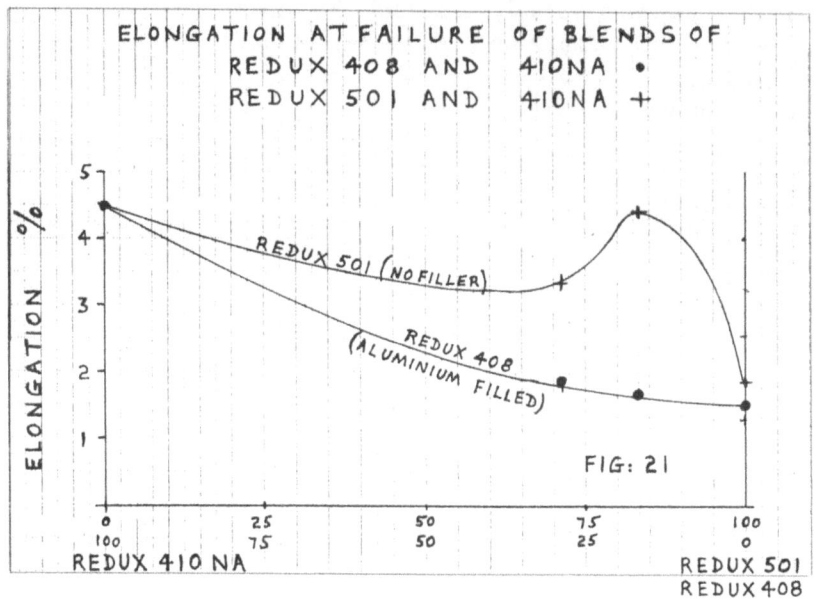

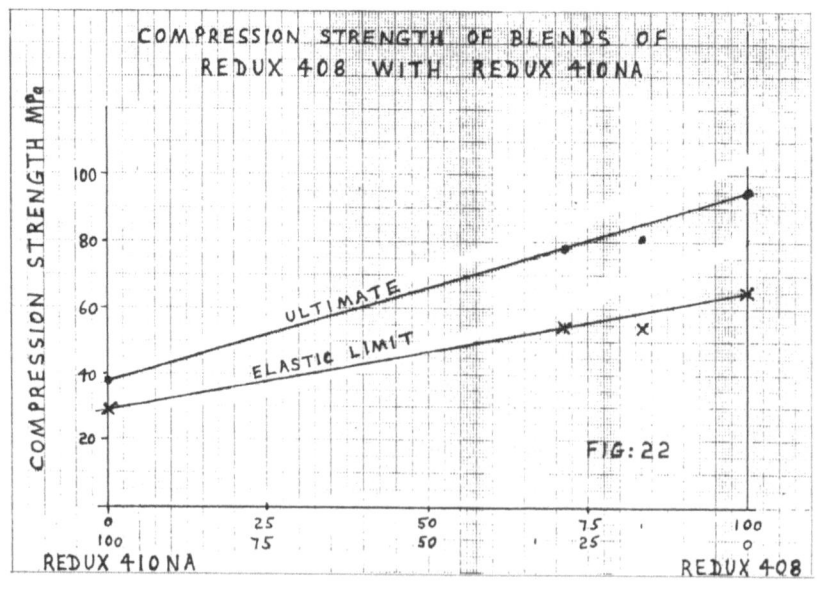

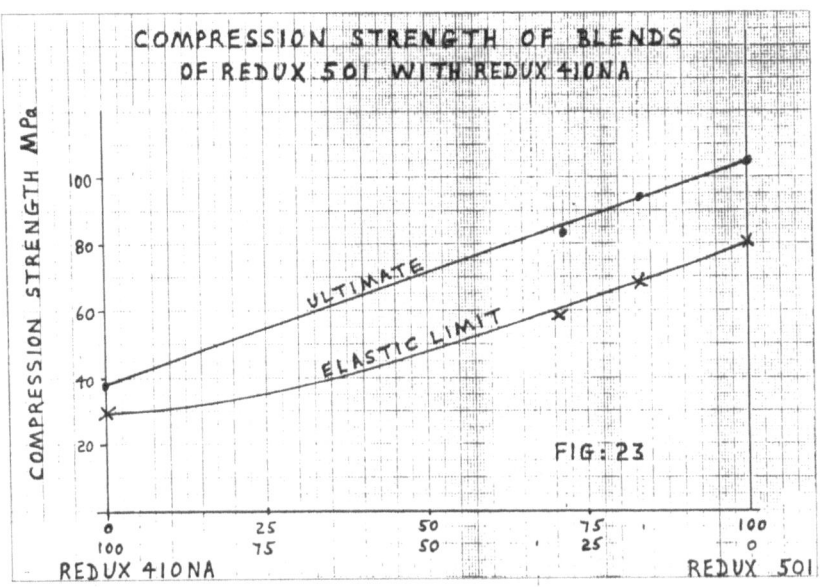

FIG: 23

Compression Strength of Blends of Redux 501 with Redux 410NA

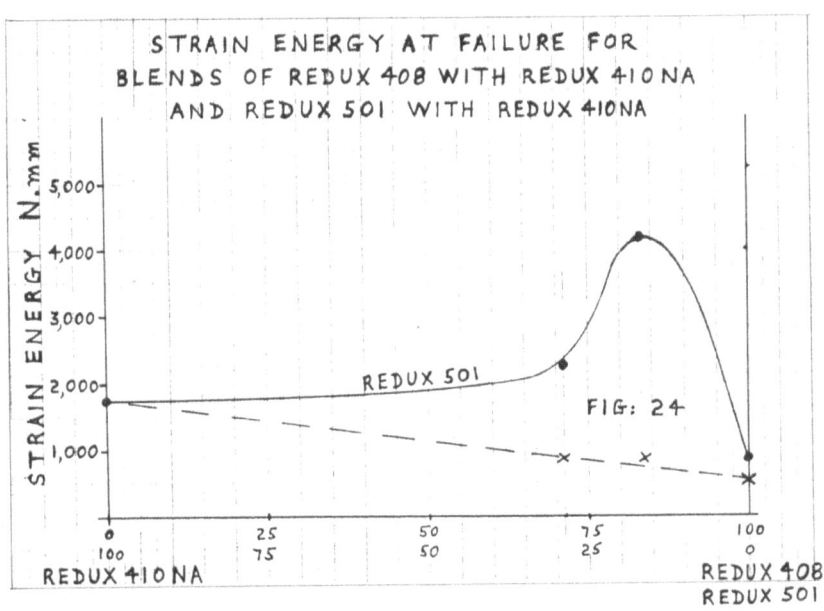

FIG: 24

Strain Energy at Failure for Blends of Redux 408 with Redux 410NA and Redux 501 with Redux 410NA

276

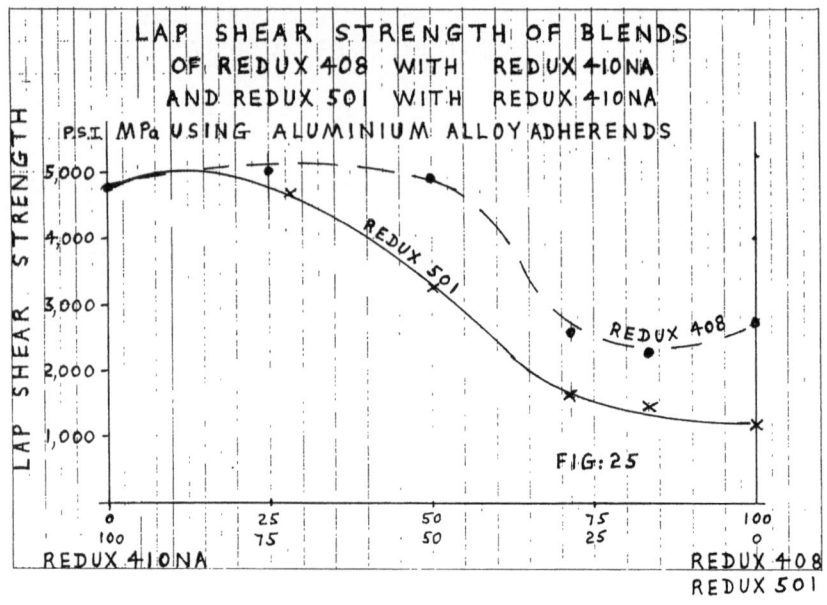

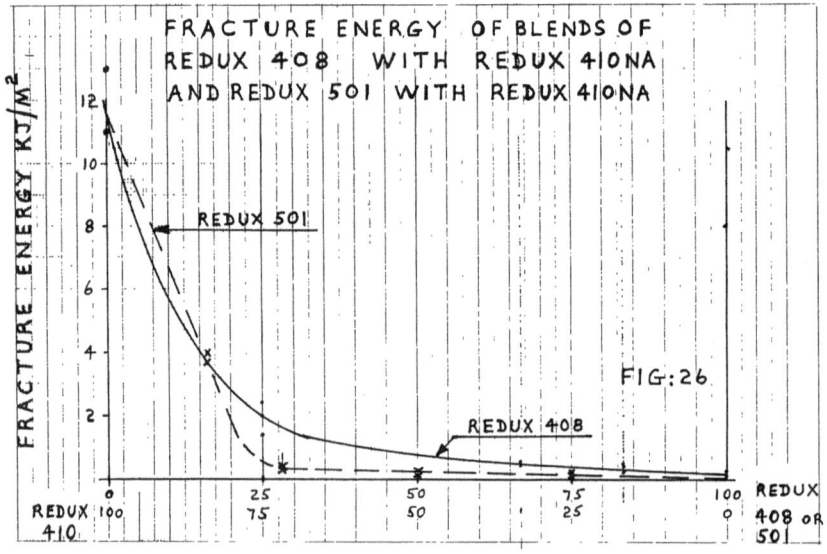

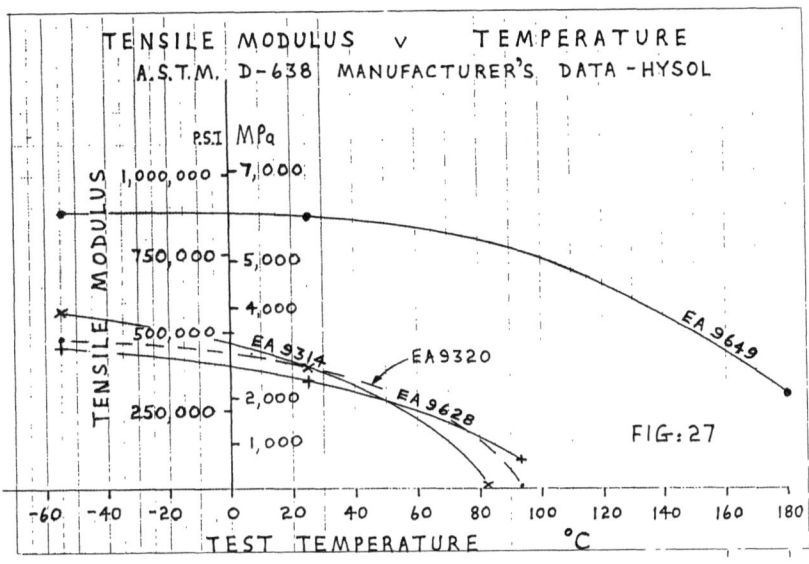

TENSILE MODULUS v TEMPERATURE
A.S.T.M. D-638 MANUFACTURER'S DATA -HYSOL

FIG: 27

16

STUDIES OF TITANIUM PRIMERS FOR ADHESION BY I.E.T.S.

K.W.ALLEN, J.E.D.SPENCER and B.O.FIELD
Adhesion Science Group, Chemistry Dept.
City University, Northampton Square,
LONDON E.C.1V OHB

1   INTRODUCTION

1.1 Inelastic Electron Tunnelling Spectroscopy

Inelastic Electron Tunnelling Spectroscopy (IETS) permits the vibrational and, in some cases, the electronic modes of adsorbed molecules to be determined with high resolution at picogram concentrations. This information can be used to define substrate interactions of importance, for example in :-

    (i)   Heterogeneous catalysis [1,2,3,4]

    (ii) Corrosion studies [5,6]

    (iii) Adhesion [7,8,9,10]

as well as other topics [11]. The technique also allows the analytical determination of very low concentrations of a wide variety of molecules in gas phase or solution. A review of the chemical applications of IETS is available [12], which affords an outline of the fundamental physical aspects involved in IETS and the various methods of tunnelling junction fabrication are considered. Briefly, however, the aluminium electrode of an $Al-AlO_x-Pb$ tunnel junction (the geometry of which is illustrated in figure 1)is vacuum evaporated onto a glass fibre support. An oxide film 30-70 Å thick is produced by corona discharge in an oxygen atmosphere. Then the species to be studied is adsorbed onto the alumina surface. The junction is completed by the vacuum evaporation of the lead electrode.

When a D.C. bias is applied, electrons may tunnel from metal 1, in this case aluminium, through the junction in two ways, illustrated in Figure 2.

    (i)   Elastically - the electrons tunnel to empty adjacent states in metal 2 without interacting with the oxide insulator or dopant molecules.

    (ii) Inelastically - some electrons (approx. 1%) tunnel across the oxide interacting with the oxide barrier and/or the dopant and lose energy characteristic of this interaction ($h\nu_0$).

278

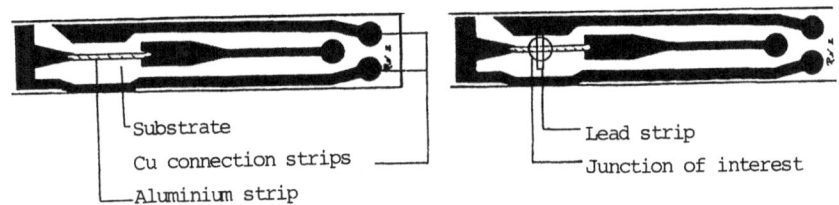

Substrate
Cu connection strips
Aluminium strip

Lead strip
Junction of interest

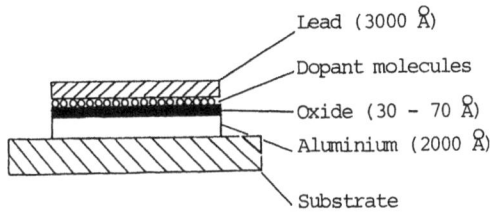

Lead (3000 Å)
Dopant molecules
Oxide (30 - 70 Å)
Aluminium (2000 Å)
Substrate

Figure 1. The Geometry of a tunnel junction.

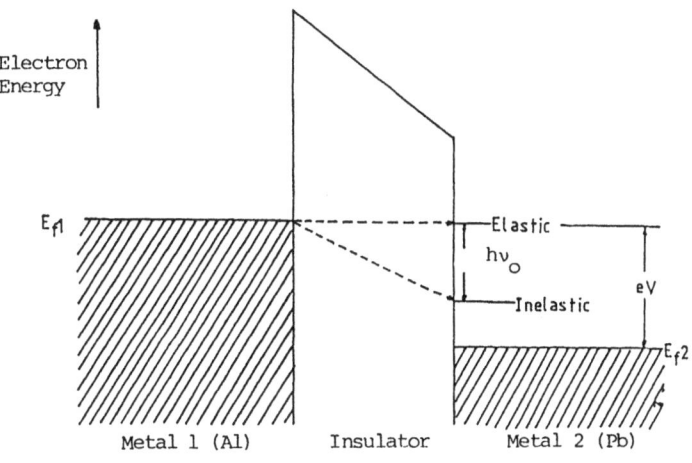

Figure 2. The tunnelling processes in a junction.

These interactions produce a non-linearity of the tunnelling current- voltage plot and may be distinguished more easily by plotting the second derivative d^2I/dV^2, see Figure 3. This is produced by applying a small A.C modulation voltage, normally less than 4mV, to the sweeping D.C. bias. A small second harmonic voltage results across the junction which is proportional to the second derivative. The signal is detected by a lock-in amplifier and the spectrum produced. Signal/noise enhancement is accomplished by taking 4000 data points 5 times and averaging them. The amplifier detects picovolts and therefore the equipment is placed in a Faraday cage and the power mains are screened to minimise both radiated and air-borne electrical noise.

Empirical evidence suggests that bonds within the adsorbed dopant species which are aligned perpendicular to the oxide surface show enhanced interaction with the tunnelling electrons when compared with the interactions due to bonds that are parallel to the surface [13,14]. This effect then not only allows observation of the dopant molecules and their interaction with the oxide barrier but also their surface orientations.

The junction is scanned at liquid helium temperatures ($<4.2°K$) to produce the tunnelling spectrum. This improves resolution in two ways :

    (i)   The lead electrode becomes superconducting.

    (ii)   Transitions between the metal fermi levels occur from energy levels close to the ground states.

Aluminium has been widely studied as a substrate for adhesive bonding and therefore IETS is especially suitable for the study of adhesion and adhesion promoters on alumina surfaces [15,16,17].

1.2 Organo-Titanium Coupling Agents

In the early 1950s there was considerable interest in the use of metal alkoxides as accelerators in the drying of paints. It was found that these accelerators also had a beneficial effect on the quality of the paint coating. It was found that cellulose ester films had better heat, solvent resistance and mechanical properties when metal alkoxides were incorporated. Also the adhesion of the paint film to substrates was improved.

One of the groups of alkoxides found to be useful were those of titanium. The lower alkoxides of titanium were found to be too reactive and caused immediate gelation. This could be moderated either by using a solution of the alkoxide in its parent alcohol or by employing a less reactive, structurally more complex, alkoxide. Chelating ligands such as 1,3 diols (e.g. octylene glycol); ketoalcohols (e.g. diacetone alcohol) or ß-diketones were found to be particularly effective in moderating the reactivity of titanium alkoxy compounds. It is believed that the metal alkoxide causes cross linking in the paint medium by reacting with hydroxyl groups so leading to polymerisation. Addition of the parent alcohol affords competition with the paint hydroxyl groups and so lowers the rate of the

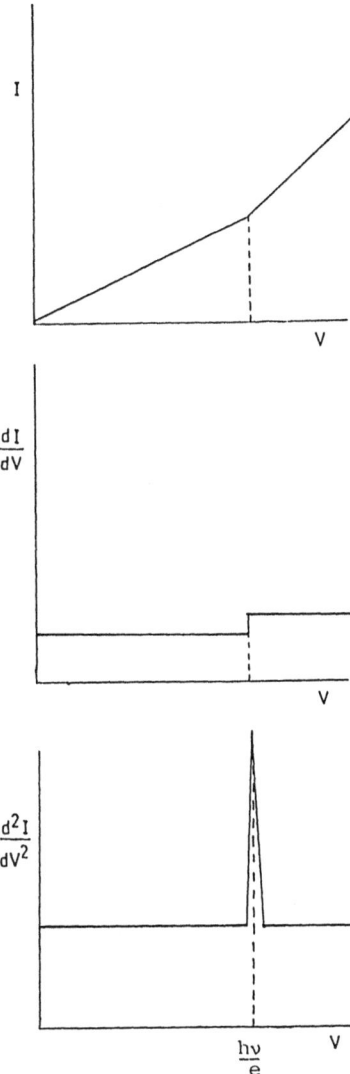

Figure 3. Current - voltage characteristic and derivatives of a tunnel junction.

crosslinking reaction. It seems probable that the ability of titanium (and zirconium) alkoxides to promote rapid drying in printing inks also makes use of the reactivity of the alkoxides as crosslinking agents. This is illustrated below.

$$\text{Polymer-OH} + \text{R-O-}\overset{\overset{\text{L}}{|}}{\underset{\underset{\text{L}}{|}}{\text{Ti}}}\text{-O-R} + \text{HO-Polymer} \longrightarrow \text{Polymer-O-}\overset{\overset{\text{L}}{|}}{\underset{\underset{\text{L}}{|}}{\text{Ti}}}\text{-O-Polymer}$$

$$+ 2\text{R-OH}$$

Where   L = chelating group
and   R = alkyl group

Three industrially important titanium chelates have been investigated. Samples of which were generously supplied to us by Dynamit Nobel Chemie: titanium acetylacetonate (**TAA**); triethanolamine titanate (**TEAT**) and octylene glycol titanate (**OGT**).

Where R = HOCH$_2$CH-CH--
               Et Pr

Titanium Acetylacetonate   **Triethanolamine Titanate**   **Octylene Glycol Titanate**

Owing to their special structure these titanium chelates are much more stable to hydrolysis than the corresponding simple alkyl compounds. The alkoxy groups still present in the chelates are shielded by the chelating groups so that their hydrolytic reactivity is inhibited. Depending on their composition, aqueous solutions of these chelates may exhibit hydrolytic stabilities ranging from a few hours to several weeks. The hydrolysis of some solutions of these titanium chelates have been investigated by IETS.

EXPERIMENTAL

The method of tunnel junction fabrication has already been outlined [12]. The dopant species was introduced into the tunnelling junction by spin doping. A drop of a dilute solution was placed onto the alumina of the tunnel junction and left for a specific period of time ($\approx$ 1 sec). The excess liquid was spun off at 3000 revs/min and the junction completed by depositing the lead by vacuum evaporation at 10^{-5} torr.

Molecular sieve dried isopropyl alcohol (**IPA**) was used as the solvent for most of the titanate solutions that were studied. It was found that IPA alone would not dope the alumina surface under the spin doping conditions outlined above and therefore interpretation of the resulting spectra assumed the absence of any peaks due to the adsorption of IPA.

2.1 Titanium acetylacetonate doped tunnel junctions

Figure 4 shows the spectrum of a junction spin doped by a 0.5% v/v TAA in anhydrous IPA solution. The main peaks of the spectrum were attributed to C-H and Ti-O vibration modes. The chelate ring modes, normally at 200 meV (1600 cm^{-1}) appeared greatly reduced indicating, by the selection rule, that the rings tend to be parallel to the oxide surface. The absence of Ti-O-Ti vibrational modes at 904 cm^{-1} indicates that, at this stage, no cross polymerisation of the titanate molecules had occurred. Figure 5 illustrates the surface orientation of TAA from the anhydrous doped solution. It shows that no solution hydrolysis had occurred. The chelate rings are parallel to the oxide surface and this implies the loss of an isopropyl unit for surface interaction.

Figure 6 is the tunnelling spectrum of 0.5% v/v TAA and 0.5% v/v H_2O simultaneously in IPA doped as outlined above. It shows several differences from the previous anhydrous case. Peaks at 1018 and 1087 cm^{-1} correspond respectively to the symmetric and antisymmetric stretching frequencies of a Ti=O bond. The frequencies of the Ti-O stretching modes at 532 and 412 cm^{-1} were found to have been reduced by 30-40 cm^{-1} in comparison to the corresponding modes in the spectrum of the anhydrous TAA solution doped tunnel junction. The C-H deformation modes were also reduced in intensity. These factors suggest that solution hydrolysis of the TAA molecule had occurred to produce the structure illustrated in Figure 7. The reduction in the intensity of the C-H deformation modes implies that both isopropyl units of the molecule have been lost in the production of the Ti=O double bond. However the relative low intensity of the chelate ring modes again suggests that the rings tend to be parallel to the oxide surface. PURI [18] has suggested this hydrolysis reaction of titanium bisacetylacetonates in moist alcohol. The absence of Ti-O-Ti modes at 904 cm^{-1} in Figure 6 implies that although hydrolysis has occurred, there is no evidence for the polymerisation of the hydrolysis products.

Figure 8 illustrates the IET spectrum of 0.5% v/v TAA in water in the absence

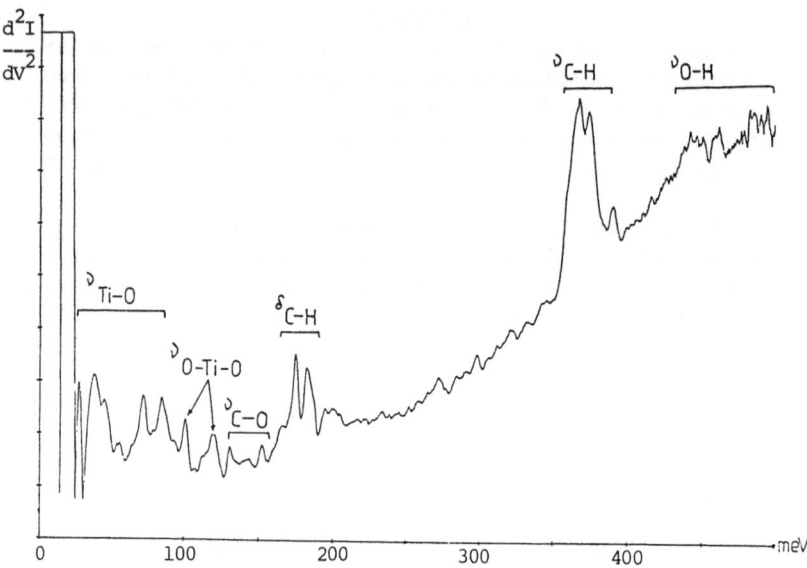

Figure 4. Tunnelling spectrum of 0.5% v/v T.A.A. in anhydrous solution

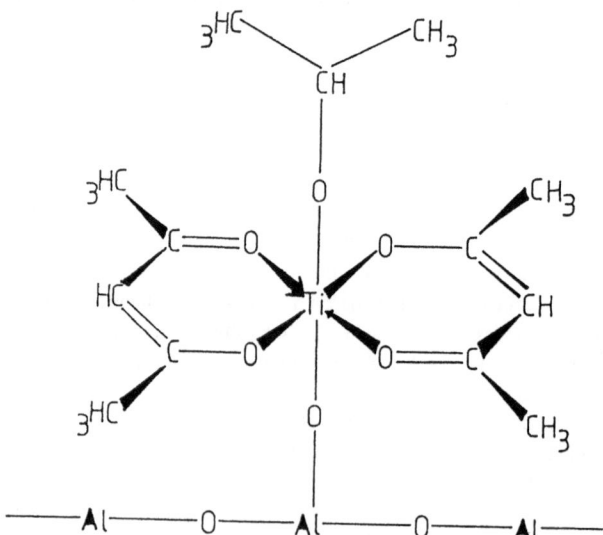

Figure 5. Proposed surface orientation of T.A.A. from

an anhydrous alcohol solution

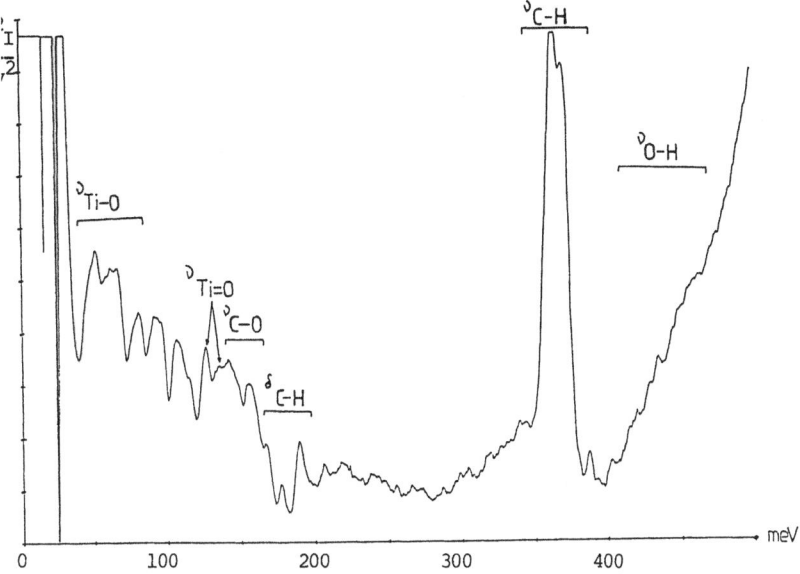

Figure 6. 0.5% v/v T.A.A. and 0.5% v/v H_2O in I.P.A.

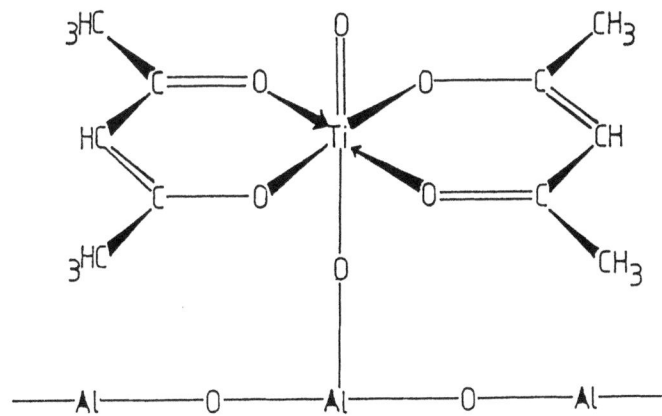

Figure 7. Proposed solution hydrolysis product from T.A.A. in

a moist alcohol solution.

of IPA. This solution was prepared 50 hours prior to doping to allow sufficient time for any solution reactions to take place. This spectrum shows, by the predominance of Ti-O vibrational modes and the absence of the C-H modes, that considerable hydrolysis of the TAA had occurred in the solution and subsequent polymerisation of the hydrolysis products taken place. Figure 9 illustrates the proposed reaction product. The low intensity of the Ti-O-Ti vibrational modes at 904 cm^{-1}, compared to that of the O-Ti-O bonds may be explained by their relative orientations on the alumina surface. The Ti-O-Ti bonds are predominantly parallel to the surface whereas the O-Ti-O bonds are perpendicular and therefore appear more intense in the tunnel spectrum.

In summary these spectra of titanium acetylacetonate indicate the solution hydrolysis scheme shown in Figure 10.

2.2 Triethanolamine titanate doped tunnel junctions

Figure 11 shows the IET spectrum of 0.5% v/v TEAT in anhydrous IPA. The presence of N-H and -NH$_2$ deformation and stretching modes at 1610 and 3300 cm^{-1} respectively suggest that reaction of the tertiary amine has occurred either in solution or on the alumina surface. The resulting N-H vibrational modes were found to occur at a slightly lower frequency than would be expected for an unperturbed amine. This is suggested to result from their interaction with the alumina surface oxygen atoms. The intensity of the Ti-O vibrational modes suggest a similar surface adsorption of TEAT to that of TAA from the equivalent anhydrous solution in IPA (see Figure 5). The absence of a C-N stretching mode in Figure 11 (1040 cm^{-1} in the i.r. spectrum of TEAT) implies that the chelate rings are again parallel to the surface. Figure 12 illustrates the proposed orientation of TEAT from this anhydrous solution on the alumina surface. The ethanol that would be produced from the hydrolysis of the tertiary amine has been shown not to "dope" under the conditions used in the production of this spectrum.

Figure 13 shows the IET spectrum of 0.5% v/v TEAT and 0.5% v/v H$_2$O simultaneously in anhydrous IPA. It indicates that, as in the case of TAA in a moist alcohol solution, hydrolysis has occurred to produce a species containing a Ti=O bond. This is further substantiated again by the reduction of the Ti-O vibrational frequencies by 20-30 cm^{-1} in comparison with the equivalent peaks in the anhydrous solution spectrum (Figure 11). The N-H and -NH$_2$ modes in the wet solution spectrum are similar in energy and intensity to those of the equivalent peaks in the anhydrous solution spectrum. This suggests that no further hydrolysis of the amine group has occurred in the wet solution. Therefore this implies that the hydrolysis of the amine group occurs as a surface induced effect. From the evidence above it would be expected that the interaction of TEAT from the moist alcohol solution would be similar to that of TAA in Figure 7 with the chelate rings parallel to the surface and a Ti=O bond perpendicular. The absence of Ti-O-Ti vibrational modes from Figure 13 indicates that although hydrolysis of

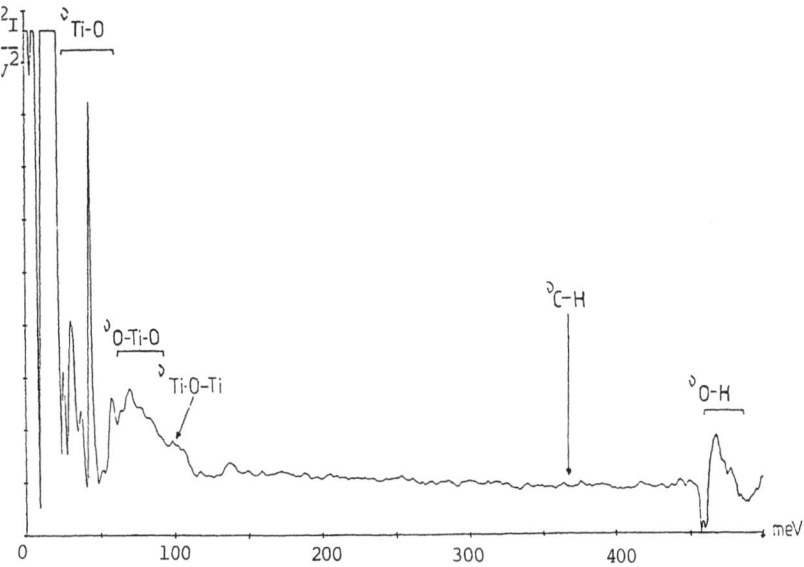

Figure 8. 0.5% v/v T.A.A. in H_2O

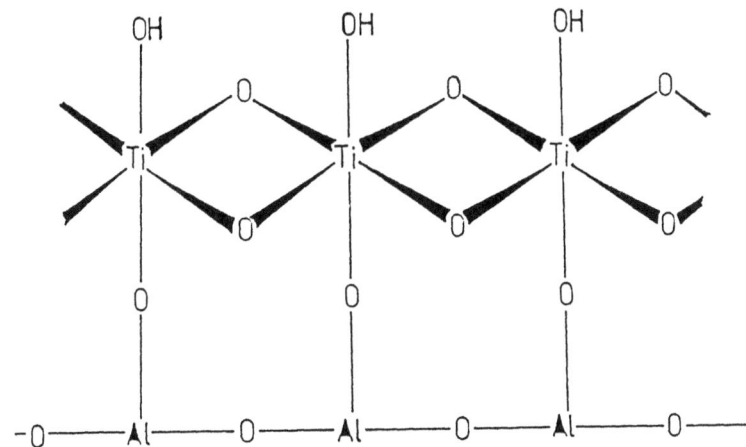

Figure 9.   Proposed surface orientation of the final hydrolysis

product from   T.A.A. in water.

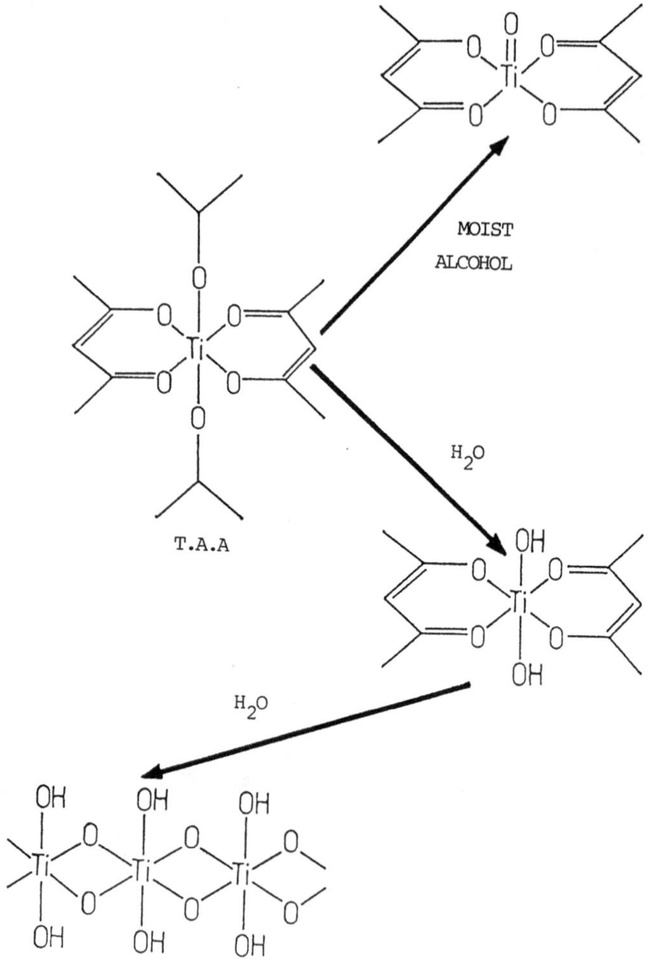

Figure 10.   Proposed solution hydrolysis scheme of T.A.A.

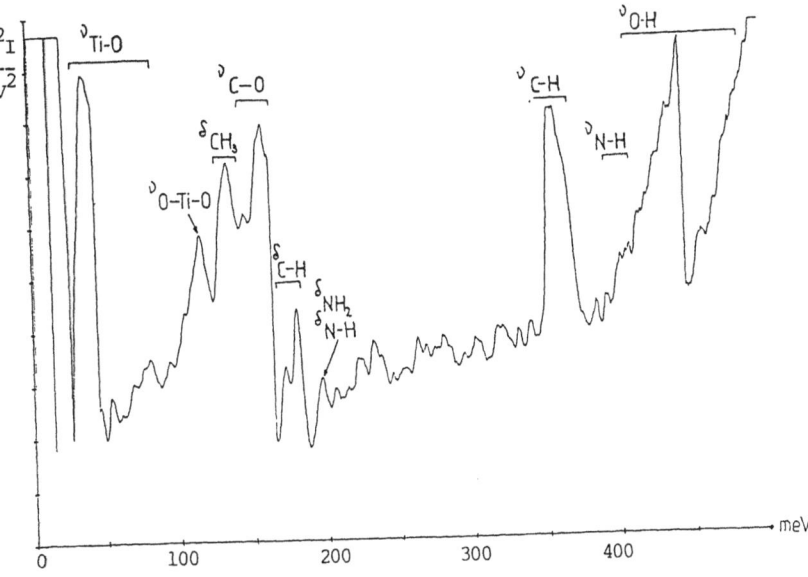

Figure 11. 0.5% v/v T.E.A.T. in anhydrous I.P.A.

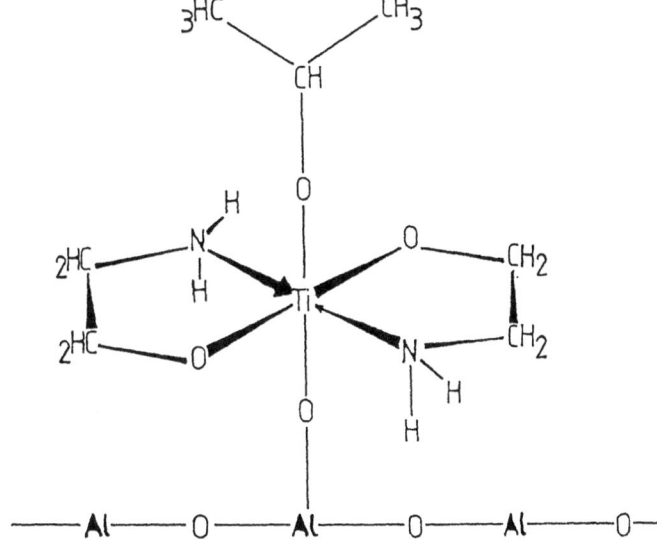

Figure 12. Proposed surface orientation of T.E.A.T from an

anhydrous alcohol solution.

TEAT has occurred, in the moist alcohol solution, no evidence is observed for polymerisation reactions. However the similarities of the initial surface interactions of TEAT and TAA and their reactions in moist alcohol solution would suggest that TEAT in aqueous solution would undergo solution hydrolysis and polymerisation as suggested for the case of TAA.

2.3 Octylene glycol titanate doped tunnel junctions.

Figure 14 shows the IET spectrum of 0.5% v/v OGT in anhydrous IPA. By comparison of the peaks present in this spectrum with the infrared spectrum of liquid OGT no distinguishing features could be found to uniquely indicate the interaction of OGT with the alumina surface. However the strength of the O-Ti-O and Ti-O modes as well as the absence of Ti-O-Ti modes suggests that the OGT molecule has not undergone any significant hydrolytic reaction in the anhydrous solution. This junction was removed from the liquid helium dewar after the spectrum in Figure 14 had been produced and exposed to wet air for 10 minutes . The reason for this was to analyse the possible effect of surface induced hydrolysis on the adsorbed surface species. The junction was then replaced in the cryogenic dewar and the resulting spectrum is shown in Figure 15. The spectrum of the exposed junction proved similar to that of the unexposed junction except that the O-H deformation and stretching modes increased in intensity and the O-Ti-O stretching mode was split. No evidence for polymerisation can be seen although some hydrolysis of the OGT may have occurred. The junction was again removed from the dewar but this time was exposed to wet air this time for 15 hours. Figure 16 shows the resulting IET spectrum. It suggests that considerable surface hydrolysis has taken place during exposure, however again the absence of Ti-O-Ti modes indicates that no polymerisation had occurred. The presence, in Figure 16 of strong C-H stretching modes may be contrasted with the ultimate solution hydrolysis product of TAA. The latter showed considerable cross- linkage but the absence of any organic moiety (see Figure 8). This suggests that the fixed Ti-centres produced by the surface hydrolysis of OGT are unable to interact to produce a polymeric Ti-O-Ti species.

CONCLUSION

Inelastic electron tunnelling spectroscopy has been used to elucidate the solution hydrolysis and polymerisation of titanium chelates and their mode of interaction with an alumina surface. This illustrates the application of IETS to adhesion science and future work in this field will permit a complete assignment of the interaction of the coupling agent with the substrates and perhaps with the paint or adhesive media. This information should allow more effective coupling agents to be designed.

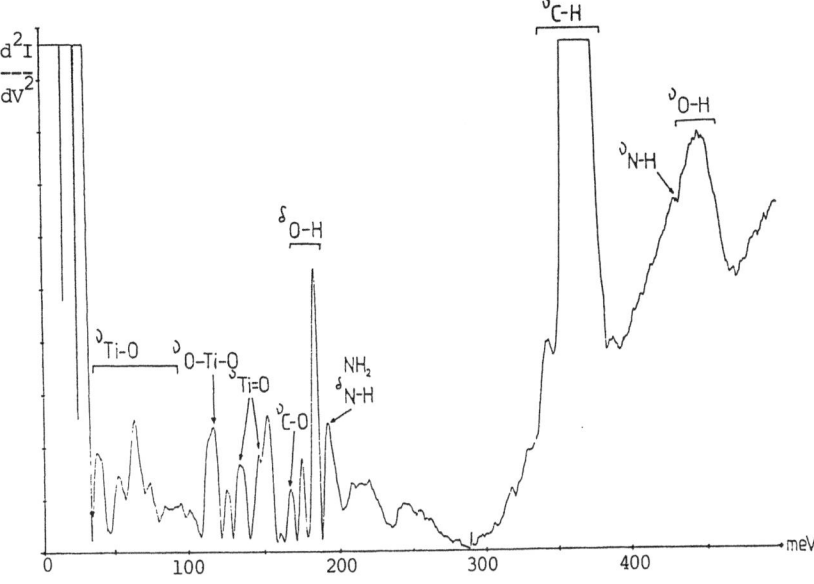

Figure 13.   0.5% v/v T.E.A.T. and 0.5% v/v H_2O in I.P.A.

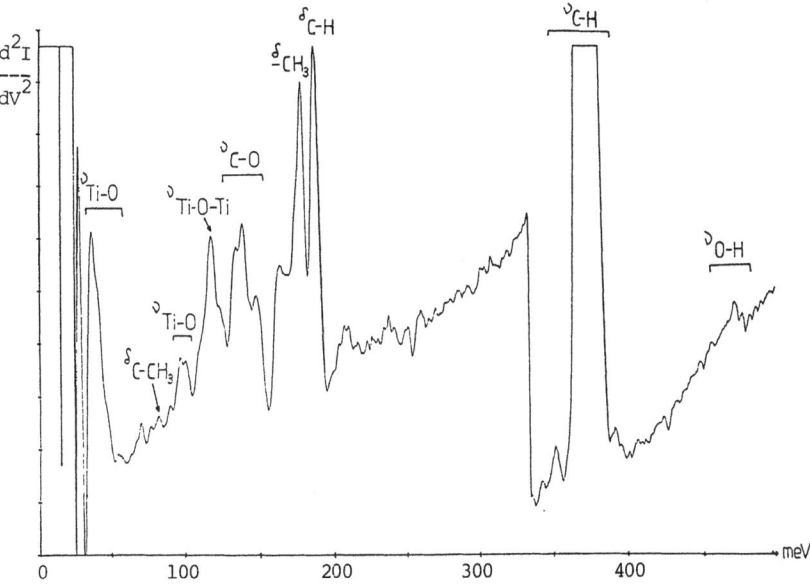

Figure 14. 0.5% v/v O.G.T. in anhydrous I.P.A.

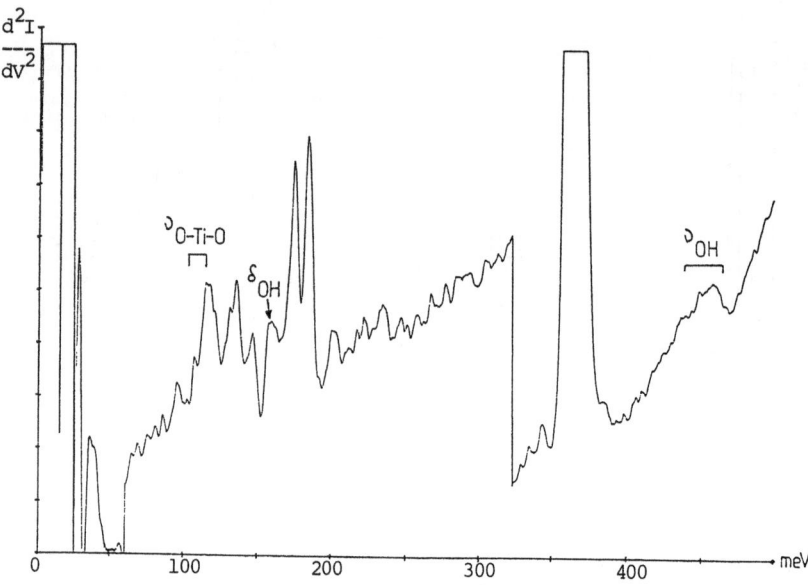

Figure 15. 0.5% v/v O.G.T exposed to air for 10 minutes.

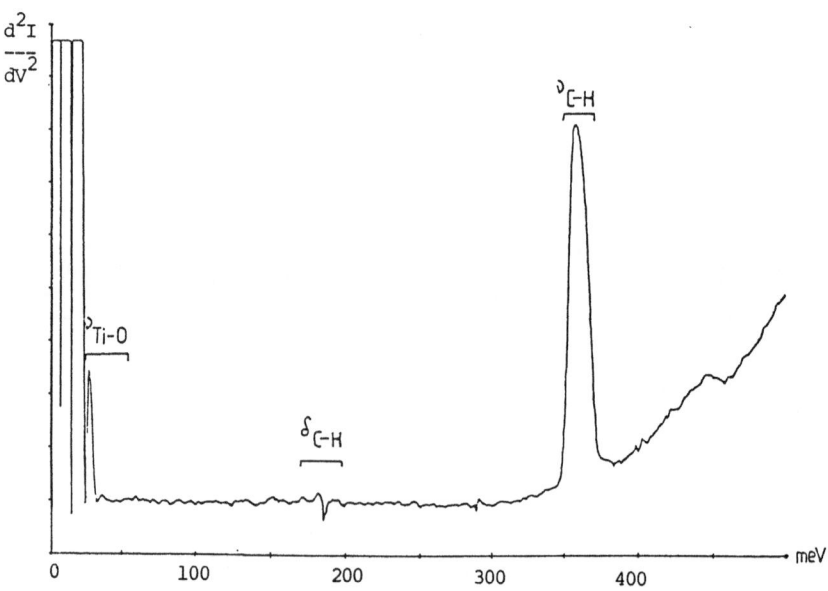

Figure 16. 0.5% v/v O.G.T. exposed to air for 15 hours.

293

REFERENCES

[1] Klein, J., Leger, A., Belin, M.,Defourneau, D. and Sangster, M.J.L., Phys.Rev.B, **7**, 2336 (1973).

[2] Field, B.O., Lewis, D.M. and Hart, R., Spectrochim. Acta, **41A**, 1069 (1985).

[3] De Cheveigne, S., Gauthier, S., Guinet, C., Lebrun, M.-M., Klein, J. and Belin, M., J.Chem.Soc.Faraday Trans. 2, **81**, 1375 (1985).

[4] Monjushiro, H., Murata, K. and Ikeda, S., Bull.Chem.Soc.Jpn., **58**, 957 (1985).

[5] Ellialtioglu, R.W., White, H.W., Godwin, L.M. and Wolfram, T., J.Chem.Phys., **72**, 5291 (1980).

[6] Ellialtioglu, R.W., White, H.W., Godwin, L.M. and Wolfram, T., J.Chem.Phys., **75**, 2432 (1981).

[7] White, H.W., Godwin, L.M. and Wolfram, T., J.Adhes., **9**, 237 (1978).

[8] Reynolds, S., Oxley, D.P. and Pritchard, R.G., Spectrochim. Acta, **38A**, 103 (1982).

[9] Comyn, J., Horley, C.C., Oxley, D.P., Pritchard, R.G. and Tegg, J.L., J.Adhes, **12**, 171 (1981).

[10] White, H.W., Godwin, L.M. and Wolfram, T., J.Adhes., **13**, 177 (1981).

[11] Hansma, P.K., (ed.), Tunneling Spectroscopy: Capabilities, Applications and New Techniques, Plenum Press, 1982.

[12] Lewis, D.M., Spencer, J.E.D. and Field, B.O., Spectrochim. Acta, **44A**, 247 (1988).

[13] Lewis, D.M. and Field, B.O., Spectrochim. Acta, **41A**, 477 (1985).

[14] Higo, M., Mizutaru, S. and Kamata, S., Bull.Chem.Soc.Jpn., **58**, 2960 (1985).

[15] Comyn, J., Adhesion 9, Elsevier Applied Science Publishers, London, ed., Allen, K.W., 1984, pp.147.

[16] Brewis, D.M., Comyn, J., Oxley, D.P., Pritchard, R.G., Reynolds, S., Werrett, C.R. and Kinlock, A.J., Surf.Interface Anal., **6**, 40 (1984).

[17] Chu, H.T., "Probing Polymer Structures", From 174th meeting of the A.C.S., ed., J.L.Koenig. p.87 (1979).

[18] Puri, D.M., Pandey, K.C. and Mehrotra, R.C., J.Less Common Metals, **4**, 393 (1962).